MOUNT WHITNEY TO YOSEMITE: THE GEOLOGY OF THE JOHN MUIR TRAIL

JAMES M. WISE

December 28, 2008

MOUNT WHITNEY TO YOSEMITE: THE GEOLOGY OF THE JOHN MUIR TRAIL

1st edition, print on demand available online at:

www.createspace.com

Copyrighted by James M. Wise, 2008.

ACKNOWLEDGEMENTS

My father Mike Wise and my friends Cheryl Hart and Lee Scotese accompanied me on several sections of the trail, providing much companionship from the otherwise solo examination of the geology of the Sierra Nevada. Mike Ressel gave helpful comments on the first draft of the guide. William Hirt provided some useful edits and suggestions for the description of the Mount Whitney Intrusive Suite. One day in the field with Tom Sisson, amid the infamous ascent of Dragon Peak in 1995, contributed excellent insight on the Onion Valley layered mafic complex. Discussions with Basil Tikoff and David Greene on the Rosy-Finch shear zone and the Sierra Crest shear zone system added an unifying element to the book. Conversations with Brian Carl clarified aspects of the Independence Dike Swarm. Kevin Mahan shared field observations on the McDoogle pluton. Discussions with Brendan McNulty, John Bartley, and Allen Glazner aided in the interpretation of pluton emplacement. Some ideas shared by Greg Stock helped in refining the sections on geomorphology of the range. My overall understanding and research of the Sierra Nevada initiated during a Masters study at the University of Nevada, Reno and my thesis advisor Rich Schweickert was instrumental for providing a foundation to examine the geologic history of the region.

The starting point for all the work in this guide came from the geologic maps produced by the United States Geological Survey, and most of the age determinations on the rocks were also produced by this agency. The U.S.G.S. has an extended history in the Sierra Nevada, from Clarence King, the first director, to the production of all the 15-minute quadrangle maps over the 1950's and 1960's when the survey was active describing the geology. The John Muir trail traverses through just a portion of the highly variable rock units studied by the U.S.G.S., although the high Sierra along the trail certainly has the best exposures.

CONTENTS
CHAPTER 1
Introduction 7
    Geography of the Sierra Nevada 9
    Climate 12
    Use of Guidebook 13
Wilderness Permits 15
Food Storage 17
Previous Geologic Investigations 18
Introduction to Geology 20
Mineral Identification 20
    Mineral Properties 21
    Rock Forming Minerals 23
Rock Classification 27
    Igneous rocks 39
    Metamorphic rocks 31
    Sedimentary rocks 33
Plate Tectonics: Driving Force of the Rock Cycle 34
Geologic Time 38
Basic Structural Geology 40
    Faults 41
    Joints 42
    Sheeting 44
    Folds 45
Plutons 46
    Features of Plutons 47
        Mafic inclusions 47
        Magmatic foliation 49
        Compositional zonation 50
        Pegmatite and aplite dikes 51
Granitic Textures and the Process of Crystallization 52
The Sierra Nevada Batholith 54
    Stoping 56
    Tectonic opening 56
    Diapirism 58
Pre-Batholithic rocks 60
Uplift of the Sierra Nevada 61
    Evidence of Modern Activity 64
    Sierra Nevada Normal Fault 66
    Timing of Uplift 70
    Phase of Earlier Uplift 73
Glacial History of the Sierra Nevada 75
    Introduction 75

Glacial features.. 76
Glacial events..80
    McGee stage 82
    Sherwin stage... 83
    Mono Basin stage 83
    Tahoe stage 84
    Tenaya stage 85
    Tioga stage... 85
    Rock Glaciers 86
Interpreting Geologic Maps 89
Overview Geology of the Sierra Nevada 92

CHAPTER 2
Mount Whitney to Forester Pass (Maps 1-5)................... 99
    Cenozoic erosion surfaces... 101
    Mount Whitney Intrusive Suite 106
    On the Origin of Schlieren 116

CHAPTER 3
Forester Pass to Glen Pass (Maps 5-7) 126
    Bubbs Creek-Cedar Grove access............................. 130
    Glacial Escape or Piracy?... 131
    Kearsarge Pass access 132

CHAPTER 4
Glen Pass to Pinchot Pass (Maps 7-10)............................ 136
    Baxter Pass access... 139
    Woods Creek-Cedar Grove access 140
    Sawmill Pass access 141
    Independence Dike Swarm 143

CHAPTER 5
Pinchot Pass to Mather Pass (Maps 10-13)..................... 153
    Taboose Pass access.. 154

CHAPTER 6
Mather Pass to Muir Pass (Maps 13-17) 162
    U-shaped Glacier Canyon Profile 167
    Bishop Pass access 172

CHAPTER 7
Muir Pass to Silver Pass (Maps 17-27) 181
    Goddard roof pendant 185
    Piute Creek-Mono Pass access.................................. 188
    Florence Lake access... 188
    Selden Pass.. 191
    Bear Creek-Lake Thomas Edison access 195

  Mono Creek-Lake Thomas Edison access 196
  Mono Pass access 196

CHAPTER 8
Silver Pass to Devils Postpile (Maps 27-32) 211
  Mount Morrison roof pendant 213
  Duck Lake Pass-Mammoth Lakes access 217
  Late Cenozoic Sierra Nevada Volcanism 223
  Devils Postpile-Mammoth Lakes access 225

CHAPTER 9
Devils Postpile to Tuolumne Meadows (Maps 32-39) 236
  Ritter Range roof pendant 237
  Shadow Lake-Mammoth Lakes access 241
  Donohue Pass 251

CHAPTER 10
Tuolumne Meadows to Yosemite (Maps 39-44) 263
  Short History of Glacial Research in Yosemite 263
  Cathedral Lakes trailhead 266
  Formation of the Tuolumne Intrusive Suite 267
  Cathedral Pass 271
  Formation of Half Dome 273
  The Giant Staircase- glacial steps 276

CHATPER 11
Yosemite Valley 287
  Joint Control of Yosemite Valley Geomorphology 287
  Mega-sheeting at Royal Arches 292
  Yosemite Valley Fill 293
    Seismic reflection data 293
    Drill hole data 295
  Extreme topography 295
  Plutonic rocks 298
  El Capitan Magma Mixing 298
  A shattered monolith 299
  Sills at the Half Dome-Kuna Crest
    Granodiorite contact, Bishop's Balcony 307
  Summary of the Late Cretaceous Magmatism 309

Appendix 1: Age Data 312
Appendix 2: Trail access 313
References 318
Index 333

CHAPTER 1
INTRODUCTION

The John Muir Trail traverses the core of the high Sierra Nevada wilderness to explore incredible exposures of glacial carved granitic and metamorphic rock, passing extreme glaciated topography of the mountain range. The Sierra Nevada is an outstanding physiographic feature of California, and contains some of the state's most beautiful landscapes. The geology of the Sierra Nevada is varied and complex. Geologists have studied this mountain range since 1860 and the research is certain to continue at a rapid pace for the next century. While the ideas and models scientists use to explain the geologic history of the Sierra Nevada may change in the future, the rocks crossed by the John Muir Trail (JMT) are fixed in terms of the human time span, exposed for everyone to study and enjoy.

The John Muir Trail, built between 1908 and 1916, covers 341 kilometers (212 miles) from the summit of Mount Whitney to the great glacial carved valley of Yosemite. Along the trail one explores impressively exposed geology, crossing several related large bodies of rock. The most notable rock assemblages are the Tuolumne Intrusive Suite, the Ritter Range pendants, the John Muir Intrusive Suite, the Mount Morrison and Goddard pendants, and the Mount Whitney Intrusive Suite. The geology of the Sierra Nevada represents time spanning from at least 550 million years ago when marine sediments were deposited, through the generation of massive igneous rocks 200 to 100 million years ago by subduction of oceanic crust beneath North America, to the most recent events of mountain uplift, volcanism, and glaciation throughout the last 5 million years. When so much rock is exposed, as along the crest of the Sierra Nevada, it is natural for one to marvel at the seemingly endless different colored masses of rock. Especially when the rock sparkles with well-formed crystals, inviting speculation on the forces that made the rocks and finally produced the mountains. Those hiking the entire JMT while reading this guide may acquire an improved understanding of the long and complex geologic history of the Sierra Nevada.

The book's main objectives are to illustrate clearly the rock formations crossed by the John Muir Trail, to point out the spectacular exposures worth taking the time to look at, and to give the general background knowledge necessary to understand what the rocks are and how they were formed. This guide covers identification techniques for rock forming minerals and rock classification, basic geologic map reading, geologic time, an introduction to plate tectonics, uplift of the range, glacial history, and other topics fundamental to origins of the rocks along the JMT. With that said, additional advanced topics are covered for which an introduction is given for in the beginning of the book and then built upon by description of the units along the JMT. Some of these topics are not covered in introductory courses in geology. Those well versed in geology are free to bypass the introduction and read between the lines. Finally, I give several short topical descriptions of interesting features of Yosemite Valley, either avoiding repetition of the classic examples or attempting to cast new light on old standing stories on the formation of Yosemite.

The geology of the John Muir trail is presented from south to north, for which I debated because ending the guide at either the summit of Mount Whitney or in Yosemite is equally spectacular. In the end, the geologic wealth of Yosemite won out for the grand finale of the guide. In terms of enjoying the view, the northbound hiker does not have to have the sun in the face every day. Comparing the initial elevation gains at the trail ends, one gains less elevation to the approach of Mount Whitney than the hike up from Happy Isles to Cathedral Pass. In other words, over the entire route (Fig. 1) the northbound hiker has more downhill because Whitney Portal is 1,318 meters (4,326 feet) higher than Happy Isles!

While the trail descriptions and maps in this book are complete enough to use for a hiker's guide on the journey, I suggest additional research into safety, equipment, wilderness permits, and survival should be done before beginning the trail. Additional useful resources include *Starr's Guide to the John Muir Trail*, and Winnett and Morey's *the John Muir Trail*. The latter guide is also formatted for the south to north trek. This guidebook does not cover the logistics of distances to hike, where to camp, or objective dangers found on the route. One should be prepared and research the potential dangers, which especially include swift cold water at the stream crossings, icy snowfields on steep slopes, and bad weather (in particular thundershowers and lightning strikes making it critical to not remain on the passes or peaks during these storms!). Not all of the rivers have bridges (and at times some of the bridges may be washed out!), and many of the stream crossings can be treacherous in the early season or in higher than normal discharge years. To hike the John Muir trail most backpackers will require 18 to 20 days, and arrange for one or more food drops to resupply. To hike the trail in one continuous trip requires planning, determination, and is a substantial accomplishment, one that will provide many years of fond remembrance. To examine and understand the geology on this adventure will enrich the experience.

GEOGRAPHY OF THE SIERRA NEVADA

The Sierra Nevada is a mountain range about 650-kilometers long and 110-kilometers wide (Fig. 2). Mount Whitney (4,416.9 m, 14,491.8'), located in the southern part of the mountains, is the highest peak in the Sierra Nevada and in the lower 48 states. From this high point the mountains gradually decrease in elevation to the north and more rapidly towards the south. The eastern side of the Sierra Nevada creates an incredible escarpment whereas the western side forms gentle forested slopes that were deeply incised by glaciers and streams. Trailheads for the John Muir Trail are at Happy Isles, in Yosemite Valley for the north end, and Whitney Portal at the south end (Fig. 3). The John Muir Trail begins at the summit of Mount Whitney, which is at least a one-day hike uphill from Whitney Portal. The eastern side of the Sierra has numerous access points climbing rapidly to reach the John Muir Trail by high passes. The west side of the Sierra Nevada provides several easy access points, and most involve longer approaches along canyon bottoms. Between the east and west side access points various sections of this guide may be used even if a continuous trek of the entire John Muir Trail is not made. The major access points of the JMT are shown in Figures 3 and 4, and are described in appendix 2 in greater detail. Entry points are identified throughout the guidebook and accompanied by GPS waypoints.

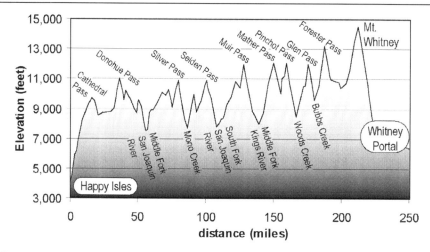

Figure 1. Graph showing the John Muir Trail as an elevation profile.

Figure 2. Location of the John Muir Trail and the Sierra Nevada batholith in California.

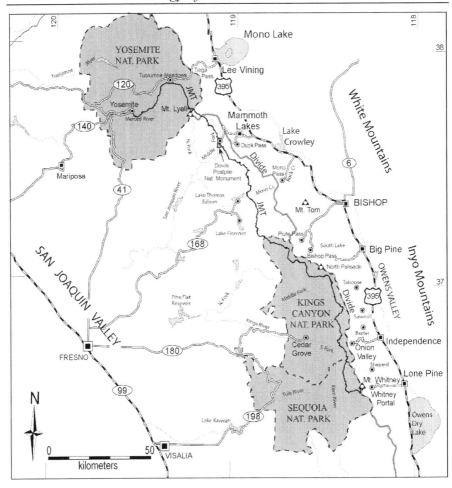

Figure 3. Location map of the John Muir Trail, major access points, highways, rivers, and outline of the national parks.

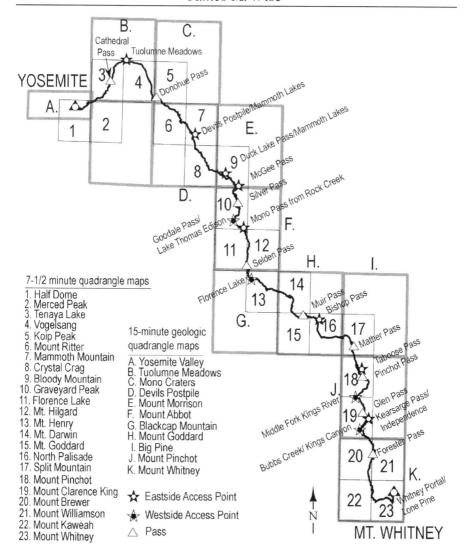

Figure 4. Index of 7.5-minute and 15-minute quadrangle maps that cover the John Muir Trail, plus west and east side access points (stars) and major passes along the JMT (triangles) are labeled.

CLIMATE

The Sierra Nevada predictably has snow in winter, unstable weather in the spring and fall, and mild conditions during the summer. Winter travel in the Sierra Nevada involves cross-country skiing and excellent knowledge of cold weather survival and orienteering skills. This is not the best time for examining the geology of the mountains for obvious reasons. Spring conditions may be excellent for backcountry travel, however, the snow level is typically substantial and spring storms can bring additional snowfall. Summer is the best time for backpacking the JMT. The weather is more stable in June, but the passes are usually full of snow. July may be the month of best overall conditions because August typically has daily afternoon thundershowers. Late summer provides the least amount of snow at the high passes, and easier fording of the major streams. In the Fall, until about October, snow is relatively scarce, except occasionally early storms in September can result in snowfall that sometimes melts off. November almost always is in winter conditions.

USE OF GUIDEBOOK

The material in the book is presented from south to north, so if a journey is made in this direction the guide is straightforward to follow. Those traveling from north to south may use the maps and geographic names listed in the index to find the corresponding text for descriptions of the geology. The guide subdivides the trail into major segments using passes along the John Muir Trail as starting and ending points, so at a minimum, the reader should be able to locate oneself into the appropriate trail segment. Additionally, if sections of the trail are hiked, the entry points provide easily located references for descriptions in the book. A list of trailheads to access the John Muir Trail is detailed in Appendix 2. A total of 44 geologic maps are in the book, each numbered, and cross-referenced in the text.

Selected GPS waypoints are listed in the text, but a GPS is not required for use of this book. I have not provided GPS points for every feature because standard map reading skills are more than adequate to locate oneself. The listed GPS waypoints in the guide were measured from maps and may include errors so that one should not expect to navigate to a feature just by the GPS read out, but these waypoints will get you close to the correct location. Locating features by use of the maps implies that one is actively looking for the comparisons and relative positions between landmarks. It may be possible to bypass the feature marked by a waypoint if paying too much attention to the GPS and not watching the character of the rock underfoot! Even so, the GPS waypoints are listed for passes (Table 1), trailheads, trail junctions, and some interesting geological features, giving another option on navigation. All previous guidebooks on the

JMT exclude the map coordinates. In an effort to make the maps more useful for those with hand held GPS units, all the geologic maps have a 2-kilometer UTM NAD27 zone 11 grid with marked coordinates. The last three digits are truncated. For example, an easting of 485,000E is simply labeled as 485. Positions can be plotted on the maps using a 1 to 25 engineering scale.

TABLE 1 TRAIL LOG OF MAJOR POINTS

	Elev. (m, ft.)	Km/miles	Waypoint (east, north)
Mount Whitney	4,416.8 (14,491.9')	0	384469E, 4048691N
Trail Crest	4,179 (13,680')	6.7, 2	384510E, 4046558N
Forester Pass	4,011 (13,160')	39.9, 24.8	377389E, 4061647N
Glen Pass	3,651 (11,978')	58.4, 36.3	374064E, 4072269N
Pinchot Pass	3,673 (12,050')	84.8, 52.7	374300E, 4088516N
Mather Pass	3,688 (12,100')	100.1, 62.2	370206E, 4099138N
Muir Pass	3,644 (11,955')	135.5, 84.2	351643E, 4108406N
Selden Pass	3,313 (10,870')	179.1, 111.3	334063E, 4128475N
Silver Pass	3,322 (10,900')	212.4, 132	330015E, 4148303N
Reds Meadow	2,316 (7,600')	250, 154.7	316746E, 4164523N
Island Pass	3,109 (10,200')	275.1, 171	306693E, 4178701N
Donohue Pass	3,368 (11,050')	282.1, 175.3	301999E, 4181473N
Tuolumne Meadows	2,606 (8,550')	307.5, 191.1	294789E, 4193588N
Cathedral Pass	2,956 (9,700')	315.7, 196.2	287549E, 4190073N
Happy Isles	1,230 (4,035')	341, 211.9	274633E, 4178845N

Throughout the text references are made to sources of information, and at first the names in parentheses may seem strange, yet it should be easy to either skim over the references paying them little notice, or to use them if more information is desired. The list of references at the back of the book provides additional resources for those having more detailed questions on the geology. Providing the complete reference is important for documenting information sources. Keep in mind that while the Sierra Nevada has had numerous studies, much remains to be examined, and the origins for many components of the range are controversial or problematic. I have included the most important studies and have not attempted to produce an absolute bibliography of Sierra Nevada geology.

Half of the maps available for the Sierra Nevada now use the metric system, whereas the other half is in the English system. Therefore, be sure to check the elevation contour intervals because they change from map to map. Contour intervals are either at 200 foot or 100 meters. The major contour lines were hand digitized, a tedious task taking more time to draft than it took to hike the JMT! The distances on the maps remain the same at 1:25,000 scale regardless of

English or metric units. The metric system is the standard measurement used in scientific studies. In the text I list both measurements in many places, but not all. If this appears schizophrenic blame it on the failure of the federal government to complete the change to the metric system, and then learn to use both units of measurement. Some conversion factors of lengths are listed below for the two systems.

>1 meter = 3.281 feet
>1 foot = 0.3048 meters
>1 kilometer = 0.6214 miles
>1 mile = 1.609 kilometers
>1 centimeters = 2.54 inches

Understanding geology takes time and thought. This guide gives descriptions of the important rock units and locations of key features. Observation on the hiker's part is required to notice much of the features described in this book. I highly recommend bringing a 10-times magnification hand lens to study the rocks. Reading the introduction to geology section is important for those with no previous geology experience. For successful use of this guide, I suggest at a minimum the guide is read during lunch times, at camp, and on the passes. It is best to make continual use of the maps and compare your location to the rock units. Actively taking observations about the rocks you are passing and using the guide as you go are necessary to conduct this geologic tour. The good side of this is stopping to read the guide and to look at the rocks is an appropriate excuse for a rest break! At times, the different igneous rocks may seem identical in memory, however, if you carry small representative chips of the units it will be clear that color, mineralogy, and textures are unique for each unit. Or one can take representative digital photographs to record a personal catalog of the rock types.

WILDERNESS PERMITS

A wilderness permit is required to hike any portion of the John Muir Trail. If the trail is being hiked in one trip send your request to the nearest permit office to the entry trailhead. Information required for a permit includes the <u>number of people in the party, entry/exit points and dates,</u> destination, and number of stock animals (if any).

Also requested are the main destination and a brief itinerary. Walk-in permits are free and account for about 40% of the trail quotas. Permit reservations can be made by mail, by phone, and more recently online. The Inyo National Forest and Yosemite National Park charges a $5 per person reservation processing fee. Kings Canyon National Park covers a $15 wilderness camping fee from May 25 through September 23. The highly trafficked and regulated Mount Whitney region calls for a $15 per person reservation fee and provides a limited application period only in February. The application fee applies to all Whitney zone entry directions (Whitney Portal and Kearsarge Pass trailheads). The permit fees and overall system seems to be continually changing, usually becoming both more restrictive and costly. Escalation or proliferation of fees to access public lands represents unfair management of our wilderness. It is a trend that needs to be checked so that our parks remain open to those of all incomes. Why should we pay to use something that belongs to us?

Checking the online webpages listed below, or searching for the park name using the word "permit." Note that web page addresses frequently change. If the below web addresses are broken then check the root for where the material has been relocated. I have not had any problems obtaining a permit on the first-come, first-served basis, especially when using the less traveled trailheads.

An "Entering Wilderness Area" sign, commonly accompanied with a list of regulations, marks most trailheads to the high Sierra. A true wilderness is wild, without any laws but those arranged by nature. The Sierra Nevada is perhaps better termed a "regulated preserve." Below is a list of the regulating agency offices where wilderness permits can be obtained.

Whitney Portal, Onion Valley, Sawmill and Taboose Passes access
Mount Whitney Ranger District
P.O. Box 8
Lone Pine, CA 93545
(760) 876-6200

Advanced reservations:
Wilderness Reservation Office
Inyo National Forest
873 N. Main St.
Bishop, CA 93514
(760) 873-2400

Devils Postpile and Duck Pass access
Inyo National Forest
Mammoth Ranger District
P.O. Box 148
Mammoth Lakes, CA 93546
(760) 924-5500
www.nps.gov/yose/wilderness/permits

Bishop and Piute Pass access
798 N. Main
Bishop, CA 93514
(760) 873-2500
www.r5.fs.fed.us/inyo

Yosemite or Tuolumne access
Wilderness Permits
P.O. Box 545
Yosemite, CA 95389
(209) 372-0740

Cedar Grove access
Kings Canyon National Park
Three Rivers, CA 93271
www.nps.gov/seki

FOOD STORAGE

Yosemite, and more recently Inyo National Forest, requires the use of bear canisters above 2,926 m (9,600'). Bear canisters cost approximately $75 or rent at about $3 per day. From my perspective, these take up more space, add weight to the pack, and are unnecessary because I have never had a bear problem during the past 20 years of trips in the Sierra Nevada. All of the times I have seen a bear, it was running away. When near or above tree line the counter balance method does not work because of short trees or lack thereof. I suggest placing food down a crack in a large boulder, using a pole or stick to recover the bag (a tied off loop on the stuff sack aids recovery). This method is safe from bears, but rodents may take tribute of the provisions. Another possibility is to hang the food from a cliff. A couple yards of 5.9 bouldering up a steep cliff to a ledge, given the right geometry of rock, can make for safe food stash. Yet another possible method, and to my knowledge untried, is to place one's food into some plastic bags and submerge them in a lake. Several imaginable pitfalls are present with this idea, but the potential is there if one wanted to experiment. The mandatory use of bear-proof containers by the government does not result in food storage being idiot proof. Those planning to send the JMT in one continuous trip will not be able to protect all food with a single container, carrying two containers is cumbersome, and hiking with empty food storage containers adds unnecessary weight to the pack. Metal bear boxes are available in the Kings and Sequoia National Parks, the locations for which are shown on the geology maps. The use of bear boxes focuses campers into one area, both concentrating human impact and attracting bears to a regular picnic ground.

PREVIOUS GEOLOGIC INVESTIGATIONS

From the gold rush of 1849 to near the turn of the previous century, geologic investigations of the Sierra Nevada were directed toward reconnaissance of the range, amounting to topographic surveying. The mountains were remote and wild, and sparsely inhabited. James Moore (2000) gives a detailed historical account of the early expeditions. Early researchers include William Blake and Lieutenant Robert Williamson during the gold rush period. The first California Geological Survey, organized by Josiah Whitney, led eventful trips that helped arrange the founding of the U.S. Geological Survey (USGS) by a volunteer of the Whitney survey. This volunteer, Clarence King, became the first director of the USGS and this in turn set into motion over a hundred years of geologic work in the Sierra Nevada. Other key people throughout the 1850 to 1900 period were George Goddard, John Muir, Joseph LeConte, Galen Clark, William Brewer, Waldemar Lindgren, Henry Turner, and Charles Hoffman, for whom many peaks in the Sierra Nevada were named.

Starting around 1900 to 1920, the USGS supported topographic surveys of 30-minute quadrangle maps by triangular methods using plane table and an alidade. This labor-intensive survey method yielded the first accurate image of the Sierra Nevada peaks, valleys, lakes, and rivers. Some early specific geologic studies were based on these maps. Adoph Knopf (1918) conducted a regional geologic investigation encompassing the eastern side of the Sierra Nevada and the Inyo Mountains. Even in 1941 the report by Mayo gave geologic descriptions in a reconnaissance format.

From 1849 to 1950, there were only about 265 published geologic reports on the Sierra Nevada. This dramatically changed in the early 1950's from the application of aerial survey methods developed during World War II that used stereo-pairs of photographs in a method called photogrammy to produce 15-minute topographic quadrangle maps. USGS started a nearly 30-year campaign of geologic mapping in the Sierra Nevada, the results of which provide the base for this book. Early workers include Paul Bateman, James Moore, Cliff Hopson, King Huber, Ronald Kistler, Jack Lockwood, and Dean Ross. These workers laid the foundation for the higher caliber academic topical studies of individual rock masses or assemblages. The advent of radiometric dating in the early 1960's and applied in the Sierra Nevada by the USGS throughout the 1970's led to increased appreciation of the batholith formation than mapping alone could provide. While mapping the quadrangles, the granitic rock masses were examined chemically, petrographically, and described the variations of density, color, and texture in individual bodies of rock. In the 1970's the USGS produced the 7.5-minute topographic quadrangle map series that completely covered the Sierra Nevada. Half of these maps were revised in the 1980's using the metric system.

Presently, more advanced methods of rock fabric measurement, called anisotropy of magnetic susceptibility (AMS), can determine the flow directions of magma before final solidification or subtle deformation fabrics imparted to the granitic rock after solidification. Refinements in radiometric dating methods, such as dating of individual zircon crystals, also yielded increased resolution on the timing of magmatism. And finally, remote sensed images from satellites give researchers new avenues on the exploration of the Sierra Nevada. Computer manipulated topographic data also is undergoing a revolution by use of digital elevation models (DEM), allowing less laborious analysis of the dimensions of the range. All topographic data developed by the USGS is available on compact disc or from various sites online. Many backpackers may be using global position systems (GPS) for navigation. This technology also allows accurate location of geologic samples, and documentation of geologic features that are not readably mapped at the 7 and 1/2-minute scale. The combination of the multiple datasets using geographic information systems (GIS), while in development for the last twenty years, has yet to be fully applied to the geology of the Sierra Nevada. The Sierra Nevada is no longer remote, one can access a wide variety of data types on the internet, and over 4,700 articles and books have been published about the geology. Despite all this information, few have walked the JMT with an eye to the ground examining the geology.

INTRODUCTION TO GEOLOGY

Several identification skills are needed to recognize the differences between the various rock types along the John Muir Trail because over 45 rock formations are exposed along the way. To the untrained eye, most rocks crossed by the trail may all seem alike. Yet with knowledge of what minerals comprise a rock and how rocks are classified, the geology exposed along the JMT has wondrous variety. To reveal the distinct features of each rock, which may at first seem subtle, the following sections of the guide are committed to explaining the minerals, rock types, geologic time, and how to read geologic maps. Reviewing this portion of the guide is essential for anyone who has not studied geology, and I suggest that if any terms in the guide become confusing, the reader should return to this section of the guide to review the topics.

MINERAL IDENTIFICATION

The earth is comprised of various rocks made from a large variety of minerals. A mineral is by definition a *naturally occurring, inorganic, structurally homogenous solid of definite chemical composition.* Anything that is man made is not considered as a mineral. Likewise, animal and plant materials do not qualify as minerals. Coal is an organic substance and therefore it is not a mineral, however, it is commonly called a mineral resource. Water is not a mineral, but ice is because it fits the above definition. Minerals and mineral groups comprising the most common building blocks of rocks are quartz, potassium feldspar, plagioclase, hornblende, pyroxene, olivine, mica and clays, and calcite. Because these minerals are so abundant it is appropriate to take a little time to learn how to identify them in a rock. To do this first we must go over the basic physical properties of a mineral, which are hardness, luster, cleavage, color, and crystal form. Other mineral properties, such as density, magnetism, radioactivity, luminescence, tenacity, refractive index, etc. are useful, but more involved to apply when in field. In practice, a ten times hand magnifying lens is helpful for examining minerals less than 5 mm across. The best approach to mineral identification is comparing the various properties and not relying on one method.

Mineral Properties

Hardness (H) is a measurement distinguishing what mineral can be scratched by another mineral. It is classified on a relative scale, called the Moh's scale of Hardness (Fig. 5). Useful in the determination of hardness are the following: a pocketknife (H = 5.5), your fingernail (H = 2.5), and a piece of quartz (H = 7.0). For example, calcite, a mineral that is commonly transparent and can be confused with quartz, has a hardness of 3. Fingernail is softer so it will not scratch this mineral. A pocketknife will easily leave a gouge in the calcite. Dragging the point of a knife across a quartz grain will not scratch the quartz. In fact, a streak of metal will be left behind and the knife blade dulled. In practice, calcite has a different luster than quartz, and has an organized fracture pattern called cleavage that aids in its identification without having to scratch the mineral. The hardest mineral, diamond, may easily scratch a pane of glass. Every mineral has a hardness and there are many minerals ranked as having the same hardness value.

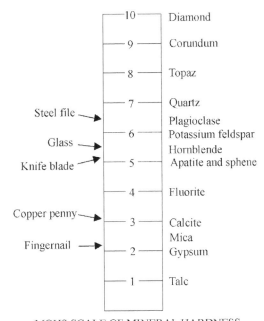

Figure 5. Relative hardness index.

Luster is the interaction of light at the surface of the mineral. Luster is described as being either metallic or non-metallic. Examples of minerals with a metallic luster are pyrite, galena, and hematite, all which may sparkle or be described as shiny. Most metallic luster minerals are opaque. Non-metallic luster has a variety of more descriptive terms that may be applied, such as glassy (for example the mineral quartz), dull, silky, resinous, and pearly. Most minerals in the rocks along the JMT have nonmetallic luster.

Cleavage is the preferred directions at which a mineral breaks. The number and directions of cleavage planes depend upon the mineral's composition. Composition determines a mineral's atomic structure and therefore the directions of weakness in the mineral. Some minerals do not have cleavage, instead they fracture, such as quartz or garnet. A crystal of calcite will break into smaller and smaller fragments having the same rhombohedral shape (a three dimensional trapezoid). Each fragment of calcite will be bound by six sides and have three cleavage directions. Quartz, lacking cleavage, will not break into repeating shapes of blocks; each direction of fracture may vary from the earlier ones. Some minerals, such as in the mica group, only have one direction of parallel cleavage.

Color is the first thing noticed about a mineral *and it must be used with caution because minerals may have more than one color.* For example garnets may be purple, red, brown, black, or green. Quartz also comes in a wide variety of colors. Nonetheless, minerals commonly have a single predominant range of color. Quartz is commonly clear to milky white, feldspars are white to buff, and hornblende and biotite are almost always black. Mineral color can change with very minor impurities, such as iron, carbon, or even water molecules, mixed into the crystal structure. Most rock forming minerals will generally have consistent color, although one should always be alert for the exceptional cases.

Crystal form provides key insight to a mineral's internal structure and is a useful property for identification. When a mineral has available open space surrounding a growing crystal the external surfaces may develop geometric patterns. If a mineral is perfectly formed it will have crystal faces and is then called **euhedral**. A mineral lacking the external faces or completed geometric form is called **anhedral**. The intermediate case, where the crystal is partially developed, is called **subhedral**. The shapes and angles between adjacent crystal faces have an immense variety. Quartz commonly forms six sided crystals that terminate in a point. Other minerals, such as pyrite or halite, form cubes. Most minerals formed in rocks where the crystal form will not be complete.

Rock Forming Minerals

Quartz is composed of silicon dioxide, SiO_2, and is generally clear and glassy in luster. It may have several other colors such as white (milky quartz), gray to brown (smoky quartz), purple (amethyst), or pink (rose quartz). Quartz has a hardness of 7.0 on the Moh's scale of hardness, thus a pocketknife (hardness of 5.5) cannot scratch quartz. When crystal faces are present on quartz it forms a six-sided geometry with an interfacial angle of 120 degrees. Quartz is one of the most common minerals at the surface of the Earth and is a major constituent of the igneous rocks in the Sierra Nevada. Quartz does not have cleavage. It breaks by concoidal fractures, a curved breakage pattern that is also seen in glass and obsidian.

Potassium feldspar, or "k-spar" for short, is actually a group of similar minerals all having the chemical composition of $KAlSi_3O_8$. The most common k-spar minerals are orthoclase, microcline, and sanidine. Orthoclase is colored opaque white, cream, or salmon pink. When complete crystals are formed they are rectangular (Fig. 6). Orthoclase is more common in the intrusive igneous rocks (rocks cooled and crystallized in the subsurface). Sanidine is a variety of potassium feldspar formed in extrusive or volcanic rocks (rocks developed on top of the land surface). Distinguishing between orthoclase and microcline generally requires use of a microscope. Orthoclase has a hardness of 6. Large crystals of potassium feldspar commonly surround other minerals as it grows, typically grains of biotite or hornblende, and these minerals are aligned parallel to the crystal faces at the time of inclusion (Fig. 7).

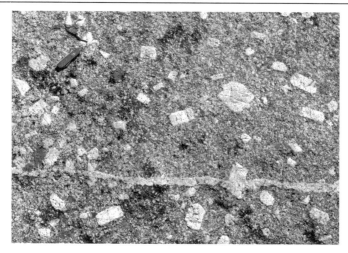

Figure 6. Rectangular potassium feldspar crystals in the Cathedral Peak Granodiorite. Swiss army knife in upper left corner for scale.

Figure 7. Photograph of a single large crystal of potassium feldspar that shows zonation and inclusions of smaller grains of black biotite arranged into rectangular rings. Penny for scale.

Plagioclase is similar to k-spar, having a slightly different chemical formula of $(Na-Ca)Al_{1-2}Si_{2-3}O_8$, which is without the element potassium. This mineral may be recognized by its elongate rectangular shape (called laths), translucent appearance (partially allowing light inside the mineral), and **twin laminae** (a structural repetition in the mineral) which in hand specimen appears as straight and fine striations similar to record grooves. Plagioclase ranges in color from clear to dark gray. It is common in both intrusive and extrusive igneous rocks. To assist in spotting the twin laminae, a key to identification, be sure to examine the hand sample from several different angles of light reflection. The striations become more apparent under certain directions of reflected light. Plagioclase may be confused with quartz, in which case remember that plagioclase contains cleavage whereas quartz does not. Note that in the above chemical formula the amount of sodium (Na) and calcium (Ca) is variable. Plagioclase comprises a series of mineral species depending on the relative amount of sodium and calcium; calcium-rich plagioclase is more common in quartz poor rock. It is common for the amount of calcium to vary in a single crystal, generally decreasing from the center outward, leading to a chemically zoned crystal.

Hornblende is an opaque black mineral composed of $(Ca,Na)_2(Mg,Fe,Al)_5 Si_6(Si,Al)_2 O_{22}(OH)_2$. It forms elongate crystals (prismatic shape) with a diamond-shaped cross sectional view (Fig. 8). Hornblende has two directions of cleavage that intersect at the angles of 120 and 60 degrees. Hardness = 5-6. Hornblende is common in igneous rocks and metamorphic rocks, and abundant in rocks along the John Muir Trail.

Biotite is a distinct mineral formed in sheets and is part of the mica group. The platy crystal shape commonly stacks together to assemble mica books. When viewed perpendicular to the mica sheets, perfectly formed biotite commonly has a hexagonal or six-sided shape. Biotite is mostly black; a similar mineral called muscovite may appear white, clear, or golden in color. Hardness = 2.5-3. Micas only have one direction of cleavage and may be flaked apart by prying with a fingernail or knife. Mica is common in both igneous and metamorphic rocks.

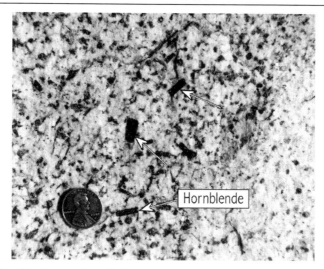

Figure 8. Photograph of hornblende crystals in the Half Dome Granodiorite. Penny for scale.

Calcite, composed of calcium carbonate ($CaCO_3$) has a hardness of 3, allowing a pocketknife to scratch it, whereas your fingernail (hardness of 2.5) will not scratch it. It is mainly white to clear in color, however it may show other colors depending on impurities. Calcite breaks by three directions of cleavage, forming rhombic-shaped cleavage fragments with an angle of 75 degrees. It is the main mineral composing limestone and marble. Calcite reacts strongly with hydrochloric acid, giving off CO_2 gas.

Olivine is a green mineral composed of $(Fe,Mg)_2SiO_4$, and it is common as crystals in basalt, a volcanic rock. Olivine has a glassy luster and no cleavage. When crystals are well formed the resulting gemstone is known as peridot. The crystal shape tends to be equant and generally forms only small grains. Hardness = 6.5-7. While the mineral is very abundant component in rocks deep in the Earth, it is relatively scare along the JMT. Olivine may be found in the lava flows near Devils Postpile.

Epidote is a dark green metamorphic mineral commonly formed in marbles by heat and fluids released from a nearby intrusion of magma. Epidote forms columnar or prismatic crystal shapes, has a glassy luster and striated crystal faces. Hardness = 6-7. Commonly joints in plutonic rock in the Sierra Nevada have coatings of light green epidote deposited from hydrothermal fluids that once flowed along the cracks.

Accessory Minerals

Accessory minerals in an igneous rock, such as granite, comprise a low percentage of the rock's total volume, generally less than 2 % of the rock, are nonetheless useful for rock identification when mapping. Most of the accessory minerals are small requiring a hand lens or a microscope to identify. Even so, a mineral called sphene sometimes forms grains large enough to identify in the field.

Sphene has a glassy luster, is honey brown, and composed of $CaTi(SiO_4)$. The crystal shape commonly appears as a flattened diamond shape. It is found as a minor accessory mineral in igneous rocks, including the Half Dome Granodiorite. Hardness = 5-5.5.

Zircon is another useful accessory mineral, though unfortunately it is commonly small and difficult to recognize without a microscope. Its hardness is 7.5 and when large enough it is used as a semiprecious stone for jewelry. Zircon is used for radiometric dating of the igneous rocks, providing most of the age data along the John Muir Trail. Zircon contains uranium, a radioactive element, which decays or breaks down at a regular rate into lead. By measuring the percentages of uranium and lead in a zircon, and knowing the rate of decay, the time when the zircon crystal formed can be calculated. The study of the numeric age of rocks is called geochronology.

ROCK CLASSIFICATION

A rock is a mixture of one or more minerals, in other words an aggregate of mineral grains. Rock types depend upon the kind of minerals present and upon the texture of the rock. Rock classification is first broadly subdivided into several major groups: **igneous**, **metamorphic**, and **sedimentary**. Most rock classifications also take into account the texture of the rock, which includes the shape and size of minerals or grains in the rock. Igneous rocks, meaning once molten, are split into two major subgroups of intrusive (plutonic) and extrusive (volcanic). The distinction between these two groups of igneous rocks is accomplished by examining the rock's texture. Rocks exposed at the Earth's surface undergo weathering through chemical and mechanical break down and result in detritus that are transported and then deposited, forming sedimentary rocks. In regions of mountain building, related to colliding tectonic plates, all types of rocks may be subjected to increased temperatures and pressures that drive physical changes in a process called metamorphism. The major rock groups of igneous, metamorphic, and sedimentary represent different states of earth materials in the upper crust, and in this dynamic Earth rocks from one group oftentimes are transformed to another group by the agents of weathering or the forces of plate tectonics, a concept known as the **rock cycle**. Most rocks, at one time or another, ultimately originated from igneous rocks. However, the rock cycle once initiated may alter a sedimentary rock directly to a metamorphic rock, melt an igneous rock to form yet another igneous rock, or any of several combinations (Fig. 9).

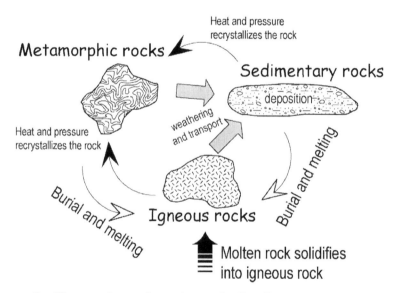

Figure 9. The rock cycle schematically illustrating the possible transformations from one major rock group to another. In-group changes may also occur, such as weathering of a sedimentary rock to deposit a second-generation sedimentary rock, or metamorphosing a metamorphic rock.

Igneous rocks

An igneous rock is any rock crystallized (formed a solid by cooling and growing crystals) from a melt. Molten rock may either crystallize underground, to become an intrusive rock, or erupt above ground, such as a lava flow, to form an extrusive rock. Extrusive and intrusive rocks may have the same exact compositions, but their textures will be different. Intrusive rocks are characterized by a coarser grain size, called a **phaneritic texture,** than that of extrusive rocks. Phaneritic refers to a rock in which the individual mineral grains can be distinguished without the aid of a microscope. Most intrusive rocks define an equigranular texture, meaning all the grains in the rock are of similar size. Extrusive rocks cool faster than intrusive rocks, not allowing the melt enough time to form large crystals. The result is an extremely fine-grained rock in which the constituent minerals cannot always be identified in a hand-sized specimen. This texture is referred to **aphanitic**. If the melt cools fast it will form a glass, called **obsidian**. Extrusive rocks may also develop larger-sized crystals surrounded by a fine-grained matrix, these larger crystals are called **phenocrysts** or **porphyry**. The overall texture of such a rock would be called **porphyritic aphanitic**. Similarly, a phaneritic intrusive rock containing a coarse matrix and even larger-sized phenocrysts defines a **porphyritic phaneritic** texture (Fig. 6). A **seriate** texture has a range of crystal sizes between the largest phenocrysts to the matrix filling crystals. If all the minerals are of similar size the rock is called **equigranular**. Grain sizes of granitic rocks are generally considered coarse if >5 mm, medium-grained for 1 – 5 mm crystal diameters, and fine-grained for crystals less than 1 mm.

Both intrusive and extrusive rocks can be separated into broad compositional groups as based on the rock chemistry. Rocks containing relatively higher percentages of calcium and potassium are called **calc-alkaline**, and represent magmatic products of continental margins such as the Sierra Nevada or the Andes of South America. Rocks poor in quartz, such as basalt that forms the oceanic plates, are called **basic** whereas those rich in quartz, commonly in continental plates, are termed **acidic**. In a similar fashion, a rock having a lot of silica in it, such as in quartz, may be called **felsic**. Felsic compositions generally give a rock an overall lighter color. In contrast, a **mafic** composition is the opposite of felsic, and it is a darker colored rock because it is relatively low in silica.

Intrusive rocks

The classification of intrusive igneous rock (phaneritic textured) composition is based upon the relative volumetric percentages, or *mode*, of quartz, potassium feldspar, and plagioclase in the rock. The percentage of these minerals plotted on a triangular composition diagram allows easy naming of the rock type (Fig. 10). Most of the intrusive rocks in the Sierra Nevada contain quartz percentages below that of granite, commonly ranging between 60 to 70% SiO_2 by weight. Note that classification of rock type by percent mineral versus percent of the chemical component varies. Recall that SiO_2 are fundamental building blocks of plagioclase, feldspar, hornblende, etc. In petrographic descriptions, other minerals contained in the rock are added to the front of the rock name as a modifier. For example, a quartz monzonite containing a small amount of sphene may be called a sphene quartz monzonite. These lesser mineral constituents are called **accessory minerals**. The "granitic" rock types (the term granitic is used loosely to refer granite, granodiorite, monzonite, etc.) are common in continental crust whereas gabbro mainly occurs in oceanic crust and is related compositionally to basalt.

Extrusive rocks

The classification of extrusive igneous rock composition is similar to that of intrusive rocks, but different names are applied to the relative amounts of quartz, potassium feldspar, and plagioclase (Fig. 11). In general, basalts are dense, quartz poor, and are colored black to dark gray; andesites are reddish to medium gray and may contain abundant hornblende crystals; dacite generally has visible quartz grains and biotite; and rhyolites are quartz-rich rocks colored light gray to white. Color in volcanic rocks can vary depending on the amount of oxidation and weathering. Because volcanic rocks are less crystalline than intrusive rocks, they tend to alter chemically at faster rates,

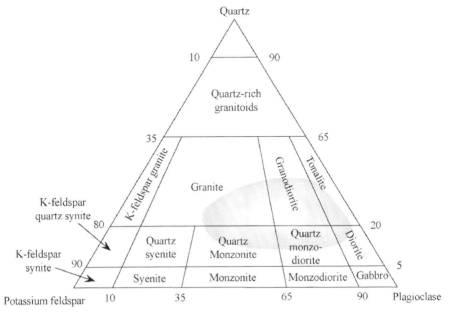

Figure 10. Intrusive rock classification based on the percentage of quartz, potassium feldspar, and plagioclase recalculated to 100 percent (diagram after Streckeisen, 1976). Most of the Sierra Nevada granitic rocks plot in the shaded region.

especially the aphanitic volcanic rocks. The fine-grained matrix, called **groundmass**, of volcanic rocks commonly includes minute crystals suspended in volcanic glass. Volcanic glass tends to be chemically unstable, quickly reacting with water to form clay minerals. Finally, depending on the composition, volcanic rocks have different temperatures of solidification from a magma. Directly measured temperatures of active basalt flows in Hawaii are generally near 1000 to 1200 degrees Celsius while molten. In contrast, rhyolite lavas have temperatures around 650 to 800 degrees Celsius.

Metamorphic rocks

A metamorphic rock is one that has undergone a physical modification as a result of temperature and pressure changes driving chemical reactions. The process of metamorphism takes place in a solid state, meaning the rock does not undergo melting. This generally restricts the temperature of metamorphism to less than 650 degrees

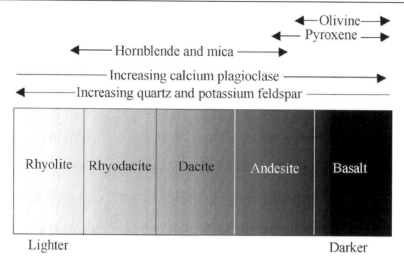

Figure 11. Generalized volcanic rock composition classification and the main types of minerals associated with each grouping.

Celsius. The two most common types of metamorphism are called **regional metamorphism** and **contact metamorphism**. Regional metamorphism recrystallizes the rock from both increased temperature and directed pressure. Regional metamorphism occurs when rocks are folded during a deformation event involving the collision of tectonic plates or continents. Contact metamorphism is mainly the result of a temperature increase. It occurs when an intrusive igneous rock comes in contact with another rock type. As the intrusion cools the heat it gives off metamorphoses, or bakes, the adjacent wall rock. Contact metamorphism generally results in a larger-grain size that shows no preferred orientation of the elongate minerals; forming a rock called *hornfels*. The two main types of metamorphic rocks along the John Muir Trail are metavolcanic and metasedimentary, both of which may be regionally and/or contact metamorphosed.

The textures of metamorphic rocks are either **foliated** or **non-foliated**. A foliated rock is characterized by preferred orientation of platy or elongate mineral, such as mica and hornblende, which results in a planar fabric. Foliated rocks are slate, schist, and gneiss, all resulting from regional metamorphism. Non-foliated rocks display a random orientation of minerals. Rocks having a non-foliated texture are marble, quartzite, skarn, and hornfels. Skarns and hornfels mainly result from contact metamorphism. Marble is a coarse-grained rock composed of calcite that was originally a limestone deposit before being metamorphosed. The original rock type before metamorphism is called a **protolith**. Quartzite develops by recrystallizing quartz sandstone, leaving little if any spaces among the quartz grains. Skarn is a special type of contact metamorphic rock formed along the margins of an intrusive rock that intruded into either limestone or marble. Skarns are typically composed of garnet, epidote, calcite, and wollastonite formed by chemical reactions from exchanged fluids with the intrusion. Hornfels are very fine-grained sediments, such as clay, silt, or shale, which are recrystallized by heat, once again related to contact metamorphism.

Sedimentary rocks

Sedimentary rocks are produced at the Earth's surface by several processes, most of which involve water. Most sedimentary rocks are a collection of grains that were once individual particles of various sizes transported by water, deposited, and cemented together to form a rock. Sedimentary rocks can be divided in terms of texture into clastic, biologic, and chemical precipitates.

Clastic rocks, such as a sandstone, are comprised of many individual particles or grains cemented together. The grains may be composed of any mineral or fragments of any kind of rock, and more common than not the main grain type is quartz, followed by potassium feldspar and plagioclase. The grain size of a clastic rock is subdivided into fine, medium, and coarse grained. If potassium feldspar grains exceed 30% of the rock, the sandstone it is called an arkose. Shale is a fine-grained clastic rock, typically dark gray to black in color, which is composed of clay and silt. Shale is commonly deposited in deep marine basins, quiet water of lakes, and the flood plains and deltas of major rivers. Conglomerate is a very coarse-grained clastic rock containing large-sized, rounded rocks called clasts that are surrounded by a sandy matrix. Conglomerates form in river channels, along beaches, and on alluvial fans. **Till** is composed of rock of all different sizes and types deposited by a glacier. Till is unsorted (meaning variable grain sizes), shows no internal layering, and composes the moraine deposits of glaciers. When it is cemented, or lithified, together it is called tillite.

The main type of biologic deposits is limestone, which is a rock composed of calcite, and forms by organic growths, such as in reefs. Deep marine deposits of planktonic life forms made of the tests or skeletons that settle to the ocean floor, form either calcareous oozes or beds of siliceous material called chert. Chemical deposits include evaporites, such as rock salt, made of the mineral halite, and gypsum formed in playas such as in the arid basins of Nevada or the extremely dry solars of northern Chile.

PLATE TECTONICS: DRIVING FORCE OF THE ROCK CYCLE

Since the Earth's accretion and consolidation, a process that combined stellar material and sorted it according to density around 4.5 billion years ago, heat has been steadily lost to space. Part of the interior's heat source is residual from the Earth's formation and additional heat is generated by ongoing radioactive decay of various elements. The Earth's internal structure, revealed from monitoring the transmission of earthquake-generated waves as they pass through the Earth, is layered and increases in density towards its center. The nucleus or core is composed of solid iron and nickel surrounded by an outer liquid core of similar material. A thick layer, composed of iron to calcium silicate rocky material, called the mantle, envelops the core. Heat is transferred, possibly by convection, from the core, through the mantle, and is lost at the surface after passing through the uppermost approximately 100-kilometer thick layer that is called the crust. The removal of heat from the Earth, and related mantle flow, drives the crust into motion that both creates new crust and destroys the old. Where new crust is formed, in areas called **spreading centers** for the separating motion of the plates, the ultramafic rock of the upper mantle, known as peridotite, is partially melted. This magma ascends to erupt from fissure system of the spreading centers, forming long mountainous ridges and valleys composed of basalt, all lying mostly beneath the ocean. At the same time, older and cooler oceanic crust is returned into the mantle along destructive plate boundaries, also called convergent boundaries. As two plates collide, one is forced beneath the other, a process that is called **subduction** (Fig. 12). Areas of subduction are responsible for some of the greatest earthquakes around the globe, including regions such as the Alaskan, Andean, and Japan trenches. Subduction of older and cooler crust back into the Earth completes the convection cell in the mantle. The whole system, known as plate tectonics at its surface expression, operates to remove heat, and in the process also reorganizes the distribution of landmass at the Earth's surface, thus in the infancy of this field of Earth science was called Continental drift. Approximately seven major plates account for most of the Earth's surface area (Fig. 13). Moreover, there are six smaller plates that may be considered as micro-plates or the tail end of an once much larger plate that has been mostly consumed by subduction. Note, that major plates include both continental and oceanic crust moving together as a single mechanical entity.

Large bodies of igneous rocks can form where two parts of the Earth's crust collide. Along the west margin of the North America continent, the Pacific oceanic crust is denser than the continental crust, so in the past when they were converging, the oceanic crust was shoved deep beneath the continent. The subducting crust, the section pulled into the Earth, will heat up if pushed deep enough into the Earth's mantle, and eventually part of the oceanic crust will melt (Fig. 12). The oceanic crust and the topping of marine sediments contain water in the pores of the rock and bound chemically in the hydrous minerals such as clay. As these rocks are pushed deep in the subduction zone the water is released. The dehydration and release of water aids in melting rocks of the hot mantle. So ultimately, the source of magma for the granitic plutons is a combination of molten oceanic crust, lower continental crust, and mantle rock.

The initial magma produced at a subduction zone from the melting of mantle rock and oceanic crust results in a melt poor in silica, such as in the compositions of basalt or gabbro. However, the magmas that formed the Sierra Nevada are granitic rocks that contain 20% or more quartz. Another intermediate process is required from the formation of a melt beneath the continental crust to the crystallization of the granite. As the melt rises upward through the crust, it differentiates, or separates into several compositions. The melt may also assimilate the continental crust as it rises, in other words, it may melt older rocks that it comes into contact with and therefore change the overall magma composition from this interaction. Commonly granitic rocks contain dark inclusions, providing direct evidence for at least two compositions of magma coexisting at depth.

All major mountain belts lie along convergent plate boundaries, such as the Andes and the Himalayas. As the plates push against one another, their leading edges thicken and deform by folding and faulting, and thus drive up the mountain belts. These belts are also called **orogens** and the process of collision and mountain building is **orogeny**. This action links into the rock cycle in two ways. First, the pressure from the colliding plates is the main force behind metamorphism. Second, elevated regions undergo accelerated erosion, providing material that is transported and then deposited into sedimentary formations. Sediments shed towards the suture of two major plates commonly becomes involved in the subduction zone and can quickly become metamorphosed, or brought to sufficient depth to be melted. Areas of subdued topography containing deformed and metamorphosed rocks can be linked to past mountain belt formation, such as in the Appalachians of eastern North America. Fundamentally, plate tectonics is the driving force behind the rock cycle, determines the surface form of the Earth, and is a process that functions over vast intervals of geologic time.

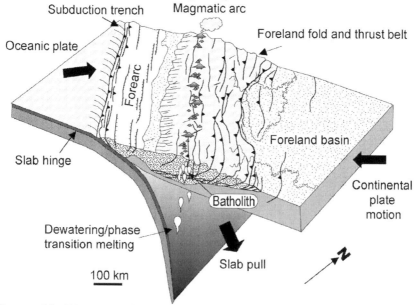

Figure 12. Diagram showing an ocean-continent convergent margin where oceanic crust composed mainly of basalt is diving beneath, a process called subduction, the less dense continental plate composed of rocks having a greater amount of quartz. At about 100 kilometers depth, the oceanic plate is heated to drive off water from metamorphic mineral reactions. This water then melts the overlying upper mantle rock to separate magma that rises in diapirs to the upper crust. The magmatic arc forms a belt of intrusions that parallels the subduction trench, and gives information about the position of subducting slab.

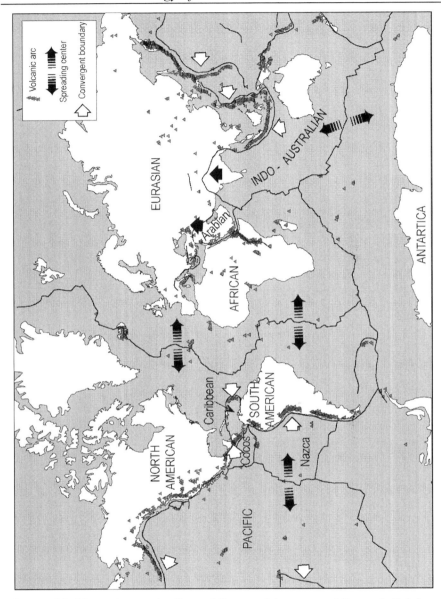

Figure 13. Mercator projection of the world map and the subdivisions of the crust into major plates. Also shown is the distribution of volcanoes.

GEOLOGIC TIME

The earth formed nearly 4.5-4.6 billion years ago, so complete geologic time spans from the present to the oldest rocks (Fig. 14). Geologic time is documented by two main methods: 1) the entire geologic time has been subdivided into various named intervals as based on the distribution of key fossils, and 2) time is numbered in million of years before present. In geology, millions of years before present is abbreviated using "Ma", especially for citing radiometrically determined numeric ages. Abbreviations for the geologic time periods are also listed in Figure 14. In addition, the Cenozoic era is subdivided into different intervals, which are from oldest to youngest the Paleocene (54.8-65 Ma), Eocene (33.7-54.8 Ma), Oligocene (23.8-33.7 Ma), Miocene (5.3-23.8 Ma), Pliocene (1.8-5.3 Ma), and Pleistocene (0.01-1.8 Ma). Sedimentary formations may be dated approximately by the types of fossils they contain, generally relying on an assemblage or group of several types of fossils. This was the first method of estimating the geologic age of rocks, but was a completely relative comparison until methods of determining the numeric or absolute age were used to calibrate the geologic time scale.

Numeric ages are calculated through the analysis of rocks containing unstable elements that systematically split their nucleus to form lighter more stable elements, a process called radioactive decay. Knowing the rate of radioactive decay for a particular element and then measuring the amount of the unstable element, called the parent, and the amount of yielded stable elements, known as the daughter products, allows the calculation of a numeric age- providing that no material is lost from the sample being dated, a concept known as a closed system. There are several methods or approaches to radiometric dating of rocks, depending on the unstable element used and the mineral analyzed. Most methods use a lab instrument called a mass spectrometer, which is a powerful magnet through which ion beam passes to a set of detectors, to measure the relative amount of particular elements in a sample. The two most useful or common sets of elements for dating are uranium-lead (U-Pb) and potassium-argon (K-Ar) pairs because of their slow decay rate. The main mineral dated by the U-Pb method uses zircon, a small accessory mineral present in most igneous rocks. The K-Ar method, and a variant approach called the $^{40}Ar/^{39}Ar$ method, is useful for dating volcanic rocks using several different minerals including plagioclase, sanidine, hornblende, and biotite. Both methods have a rigorous set of statistical criteria that must be met in order for the calculated age to be considered reliable.

The oldest radiometrically dated Earth material is 4.404 billion years old (Ga) zircon grains contained in a sandstone from Australia (Wilde and others, 2001). The oldest data material is from the meteorites, such as the Canyon Diablo and Vaca Muerta meteorites, isotopically dated at about 4.56 Ga (Chen and Tilton, 1976; Manhes and others, 1986). The ages of lunar samples from the Mare basalts range from 3.96 to 3.19 Ga (Taylor, 1975). The oldest known fossil, 3.465 Ga, is cyanobacteria from the Apex Chert of Australia (Schopf, 1993). In comparison, the oldest rocks of the Sierra Nevada are possibly Cambrian (~550 Ma). These are metamorphosed sedimentary rocks in the Mount Morrison and Big Pine pendants. Most of the granitic rocks in the Sierra Nevada are Cretaceous (80 to 105 Ma), however, Jurassic and Triassic plutons also comprise about a third of the batholith. The youngest igneous deposits are the volcanic rocks of the Devils Postpile area. The most recent events include glaciation in the Pleistocene (<1.6 Ma) and continued tectonic uplift of the range, forming the eastern escarpment and massive alluvial fans lining the base of the range in Owens Valley.

For those not accustomed to thinking about such large time spans, take heart, most geologists find boggling the amount of time addressed in astronomy. Millions of years are just a brief moment in the universe and one second is a protracted time on the subatomic scale. On Earth, cyano-bacteria and green algae has been around for over 3.5 billion years whereas the oldest hominid fossils only date back to about 6-7 million years (Vignaud and others, 2002). Instead of trying to think about how much time is represented by 80 Ma, it is better to focus on where this time is relative to other events. For example, this rock is older than event A or younger than event B for a given sequence or history of geologic events.

The relative dating between different rock units can also be constrained by using the concept of crosscutting relationships and the nature of geologic contacts. The law of superposition, developed by Nicolaus Steno in 1669, requires that in sedimentary rocks the bed deposited on top is younger than the underlying beds. Most commonly, sedimentary beds are deposited in a close to horizontal position, the principle of original horizontality. In crosscutting relationships the rock bodies that are cut or truncated must be older.

This is especially useful when examining faults or intrusive rocks. Both sedimentary and igneous rocks may incorporate fragments of older rocks, called inclusions, which is useful for determining the relative age between rock units.

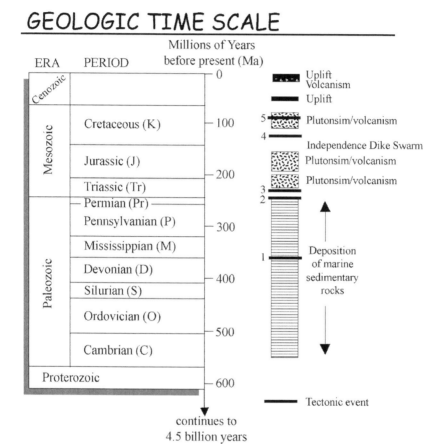

Figure 14. Simplified geologic time scale and major geologic events for the Sierra Nevada.

BASIC STRUCTURAL GEOLOGY

The Earth is very ordered and the branch of geosciences that study the architecture of rocks is called structural geology. Rocks form related packages that are marked by definite boundaries, some depositional and others produced by deformation. Horizontal layered sedimentary rocks define a structural pattern because the beds occupy tabular volumes of rock that are oriented in space. The beds are horizontal from the main force, gravity, being oriented perpendicular to the Earth's surface. Therefore, structural geology also strives to understand the forces that develop geometric patterns in rocks. Cases where the main force, called **stress**, is not vertical happens at collisions between continents and where there was folding of once horizontal layers. Structural geology has numerous subfields or topics, of which faults and folds are the most important for now to better understand the large geometric patterns of rock masses and the smaller scale, related, textures of the rocks.

During deformation rocks behave in several fundamental styles, and the study of this physical behavior is called rock mechanics. When rocks are relatively cold they will fracture and fault, or buckle (elastic behavior), and if they are hotter they will flow like honey (viscous or plastic behavior). Both mechanical styles operate simultaneously in the crust because with depth the temperature and pressure increases from the mass of the overlying rock. The types of structures observed in rocks depend on these temperatures and pressures, and on the composition of the rock. Most structures we see at the Earth's surface are from the rocks behaving as an elastic material. If you take a hammer and strike a glacier polished slab of granite, the hammer will bounce back into the air and the granite, apart from surface scratches, will be the same afterwards. The hammer rebounds into the air because the energy of the blow was absorbed elastically by the granite and by the metal of the hammer and then released to throw the hammer back. If a strong enough force is applied to a rock it can exceed the strength of the rock, causing it to break, or in other words, the rock undergoes permanent deformation.

Faults

Rocks breaking under geological force, called stress, create fractures, faults, and folds. Faults and fractures form in specific organized directions with respect to the applied direction of stress and have been classified into three basic types by Anderson (1951) as normal, reverse, and strike-slip faults (Fig. 15), each separated on the basis of the relative movement direction across the fault. Stress is a vector quantity of force acting upon an area, measured in kilobars or Mega Pascals, that in three dimensions can be summarized using three orthogonal axes of maximum, intermediate, and minimum magnitudes treated mathematically as tensors (matrix math). Note that a fracture or joint is a discontinuity or crack in the rock where the opposite sides of the crack have separated from one another, but have not moved laterally. In contrast, a fault always has some offset or displacement between the opposite sides. The amount of movement across a fault is called either the net slip or offset. For faults to form, the stress in the crust must have unequal magnitudes, unlike hydrostatic conditions where stress is distributed equally in all directions. Once a fracture or fault forms it self-modifies the surrounding stress field to concentrate it near the fault tip. The study of geologic stress is important in understanding the driving forces of plate tectonics and the mechanics of earthquakes.

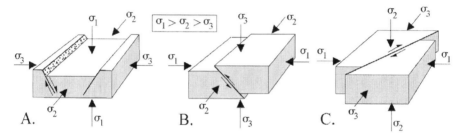

Figure 15. Anderson (1951) model of fault directions with respect to the main geologic force called the principal stress axis or sigma 1 (σ_1), and two other axes of intermediate and least principal stress directions. A. Normal fault, such as along the east side of the Sierra Nevada. B. Reverse or thrust faults common in the Himalayas. C. Strike-slip fault, in this case right-lateral slip similar to the San Andreas fault in California.

Joints are systematic planar, and generally parallel, cracks in a rock mass, and they commonly occur in several different sets of directions. Joints are a result of pressure release from the unloading of the overburden of rock, from contraction accompanying cooling of plutons, and by tectonic deformation. Overburden is the thickness of rock once above a location in the crust. Joints also form by the cooling of an intrusion of granitic rock that causes the material to contract. Joints or fractures are a response to stress in a rock, regardless of the above-mentioned driving conditions. The cracks form parallel to the maximum compressive stress direction and perpendicular to the direction of maximum tension (Fig. 16). When no lateral motion has occurred, or in other words, the fracture opens only perpendicular to the walls of the crack, the structure is called purely dilational or a Mode 1 crack (Pollard and Aydin, 1988). Cracks function to redirect or focus stress in a rock mass to the edge or tip of the fracture. Crack growth or propagation happens at various rates and may have complex interactions between adjacent cracks. Generally long cracks form by a combination of lengthening a single crack and linkage with other adjacent cracks, which sometimes can result in curvi-planar fractures or irregular jogs.

Joints play a significant role in the development of landforms of the Sierra Nevada. Most joint sets form long parallel cracks in granitic rock (Fig. 17). Glaciers and streams will remove joint bounded blocks, and both commonly form parallel to the prominent joint directions. Ice prying apart the rock (frost wedging) takes advantage of joints in breaking the rock. Commonly joints may be traced out on maps for hundreds of meters to kilometers. Numerous studies examined the processes of creating joints and why joints develop various spacings, lengths, and interactions. Overall, the regional joint patterns along the entire crest of the Sierra Nevada have not been examined any formal geologic study.

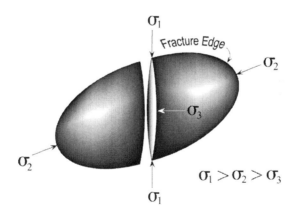

Figure 16. Relationship between compressional stress and the orientation of an opening mode crack. Walls of the crack open towards the direction of the least amount of stress, the sigma 3 direction (σ_3). Magnitudes of the three principal orthogonal stress directions are different and produce an elliptical-shaped crack. Scale of crack perpendicular dilation exaggerated with respect to the crack length and width.

Figure 17. Photograph of steeply dipping, very continuous and relatively close-spaced joint set in granitic rock on the south canyon wall of Lee Vining Creek.

Sheeting is a process responsible for breaking the granitic rock into thin slabs that parallel the surface topography (Fig. 18). It results from the immense weight of the overlying material being removed from above the granitic plutons. This removal of pressure allows the granite to expand, forming the onion skinned layering present on most domes, especially the domes in the Tuolumne Meadows area. When the sheets break away they commonly form a staircase like pattern and litter the base of the dome with piles of blocky talus. The process of slabs breaking away to form a dome or cliff is called exfoliation. Exfoliation sheets differ from joint sets by having curvi-planar forms that mimic topography. The cracks that bound the exfoliation sheets respond to stress similarly as joints, but form curved fractures because the near surface stress is non-uniform.

Figure 18. View of the north face of Fairview Dome, Tuolumne Meadows, showing rounded form and multiple massive exfoliation flakes. The dome is approximately 600 meters high.

FOLDS

Commonly layered rocks undergoing compression not only fracture, but they also fold into a variety of shapes. This is most true for originally layered rocks caught between colliding tectonic plates, such as in the Andes or the Himalayas. Along the John Muir Trail some of the layered metamorphosed volcanic and sedimentary rocks record folds. The most important parts of folds are diagrammed in Figure 19, showing an **anticline** and a **syncline**. An anticline has older rocks located in the core of the fold whereas a syncline envelops younger rocks at its center. The folded layers define a place of maximum curvature called a **fold hinge line**, which can be imagined to accommodate the bending of the layers like the motion of a door hinge. With multiple folded layers, a series of fold hinge lines describe each folded layer and a surface drawn through these lines define a **fold hinge surface**. The straighter flanks of the folds are called the **limbs**. Taking these basic forms, additional descriptors can be applied to give more information about the shape of the fold, especially the angle between the two limbs, called the interlimb angle, and type of curvature at the hinge. Additionally, orientation data on the fold hinge line and hinge surface is useful for understanding the position of the structures in three dimensions. The shapes of folds can also be generally related to the driving forces that deformed the rock. For this the hinge line represents the intermediate stress axis, sigma 2, the maximum stress axis lies perpendicular to sigma 2 and commonly in a horizontal plane, and the minimum stress is in the direction to which the rocks escapes towards lying in the plane of the hinge surface. Analysis of fold shapes and orientation can help to understand the direction of past-collided tectonic plates. However, caution must also be exercised because subsequent geologic events may rotate or reorient the folds to a new direction.

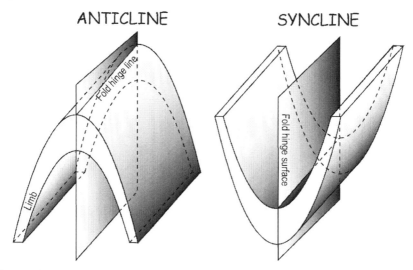

Figure 19. Basic fold types.

PLUTONS

A pluton is formed from a body of magma, a mixture of melted rock and crystals, that ascended through the Earth's crust from buoyancy much like the rising motion of colored blobs in a lava lamp (a process called diapirism). A pluton stops rising when it reaches a level where the rocks surrounding it are of a similar density. Upon halting, the pluton cools and crystallizes, forming an intrusive igneous rock, such as granite or diorite. Plutons generally are elliptical in map view and are several to many kilometers in diameter (Fig. 45). The size and shape of plutons varies considerably. Even so, the Sierra Nevada plutons commonly form elongate shapes trending to the northwest. The present erosion level exposes the plutons allowing us to see for the most part the contacts of the plutons as being steep (Fig. 20). When several plutons form side by side, forming a composite of plutons, it is called a batholith.

An ongoing field of research is pluton emplacement: how a pluton is intruded into a batholith. Questions remain about what percentage of a pluton is molten at any one time, and how mobile the magma is for different amounts of phenocrysts and melt. Recent work has shown the affects of plate tectonics on pluton emplacement. The plutonic suites crossed by the JMT record evidence for right-lateral strike-slip shearing throughout emplacement. Picture a marshmallow between your hands held straight out in front of you, keeping your hands vertical and parallel to one another. Now, press against the marshmallow while moving your right hand towards your body and your left hand away from you-- this shearing motion on the marshmallow is called right-lateral slip. Note that the marshmallow will rotate with a clockwise sense of direction.

Plutons commonly form by mixing of at least two magmas of different compositions, probably accompanied with thermal convection in the magma chamber, and repeated intrusion of melt into the chamber. For example, if a large magma chamber is occupied by a felsic melt, such as a quartz monzonite, and then intruded by more mafic material, such as diorite, the diorite may become fragmented or disrupted and then stirred into the large body of magma. However, diorite has a higher melting temperature, so commonly it is not completely mixed into the quartz monzonite. Evidence for magma mixing is abundant in the rocks along the JMT.

FEATURES OF PLUTONS

Mafic inclusions are dark colored spot-like shapes or blobs composed of diorite found in many Sierra Nevada plutons (Fig. 21). The dark inclusions have long been of interest to those studying granitic bodies. The inclusions are three dimensional in shape, defining forms that are rounded, angular, or stretched out into lenses. Mafic inclusions represent a melt of different composition intruded into the pluton while it was still molten and then mixed into the more felsic melt of the pluton (Pitcher, 1991). The inclusions sometimes show sharp boundaries or contacts with the surrounding host rock, or

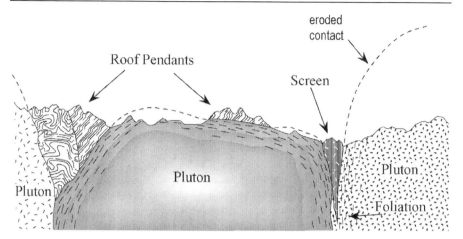

Figure 20. Schematic cross section showing steep pluton contacts marked by parallel foliation. On top of and between the plutons are remnants of the country rock, called roof pendants or screens that are generally composed of metamorphic rock, or older intrusive rocks.

they may exhibit gradational contacts and embayed margins suggesting they were melting and being assimilated into the surrounding more felsic magma. Mafic inclusions may originate by diorite being injected into a magma chamber as a dike-like body and then subsequently the dike is mechanically broken up into various sized pieces that are initially angular and then progressive become more rounded as they are stirred into the host magma. Some inclusions may be formed by disruption of accumulated mafic minerals at the floor of the magma chamber, and then the fragments become stirred into higher levels of the pluton. Minerals in the mafic inclusions commonly have chemical similarities with those in the matrix hosting the inclusion. Pitcher (1997) gives more detailed account on the formation of syn-plutonic mafic inclusions, and the development of plutons and batholith in general. Other types of inclusions are commonly present in granitic rock, known as **xenoliths** (fragments of older rock) and **restite** material (unmelted rock and minerals from the source region where the granitic rock was melted). Knopf (1918) used the near complete mixing of xenolith mineral fragments into the host magma to explain the streaky texture of concentrated mafic minerals known as **schlieren**, however, I prefer the more restricted use of schlieren as a magmatic product as discussed by Bateman (1992). The topic of schlieren is described in the Whitney to Forester Pass chapter. In general, these latter features are not immediately observable along the JMT, although xenoliths are certainly known from the contact areas with roof pendants, and some may argue that the mafic inclusions may be restite. Additional interesting references covering a wider range of inclusion types can be found in Didier (1973), Blandy and Sparks (1992), Cobbing (2000), and Collins and others (2000).

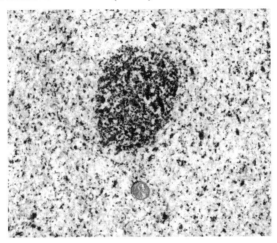

Figure 21. Photograph of a very coarse-grained mafic inclusion in the Half Dome Granodiorite. Large black phenocrysts are mainly hornblende. Penny for scale. In detail, some of the hornblende crystals in the matrix granodiorite are aligned parallel to the outer contact of the mafic inclusion, perhaps adhering to the clast to form a magmatic "snowball".

Magmatic foliation is the parallel alignment of crystals formed while the rock was partially molten. Magmatic foliation gives clues about how the once molten rock in the pluton was moving, because elongate minerals will be generally lie parallel with the flow direction. In outcrop it may be recognized by the similar alignment of mica plates and needles of hornblende, or by the elongation direction of deformed mafic inclusions (Fig. 22). Foliation commonly parallels the margins of a pluton, becoming more clearly defined closer to the zone of the pluton-wall rock contact. Most plutons are foliated, although, the fabric can be subtle. Foliation in granitic rocks can also form by deformation after the rock was cooled into a solid, and distinguishing this deformation fabric from that of magmatic foliation may be difficult in the field (e.g., Paterson and others, 1989).

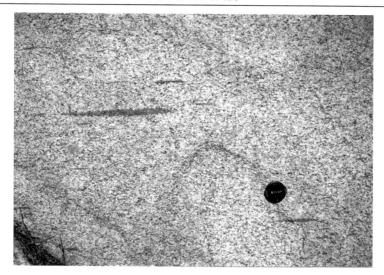

Figure 22. Well foliated granitic rock exposed near the top of Yosemite Falls. Foliation is defined by mineral alignment and highly flattened mafic inclusions. Lens cap for scale.

Compositional zonation of a pluton is a common pattern observed in most of the intrusive bodies of the Sierra Nevada. These zoned plutons include a range of compositions that are more mafic in the outer portions (having less quartz) than the interiors. If one were hiking across a compositionally zoned pluton, one would observe an increase in the minerals quartz and potassium feldspar near the center of the pluton. The variation of composition within a single pluton results from a process called crystal fractionation. Starting with a completely molten pluton, or magma chamber, as the melt cools the minerals will precipitate into solids. The first minerals to crystallize are relatively lacking in silica and water, such as olivine and pyroxene. Removal of these minerals enriches the remaining melt in whatever elements the early-crystallized solids did not use. In this way, as a pluton cools and solidifies, it becomes progressively more concentrated in quartz, potassium feldspar, muscovite, and water. The last material to solidify commonly intrudes into fractures and crystallize to form pegmatite and aplite dikes, both are water and quartz-rich fluids before solidification. In a closed system containing elements for all the rock forming minerals, the order of mineral crystallization is called *Bowen's reaction series*, after the laboratory work of Bowen (1928; Fig. 23). Plutons of the Sierra Nevada crystallized in an open-system where multiple re-injections of parent melt interplayed with the fractional crystallization process.

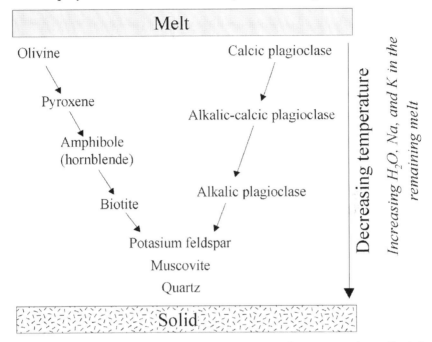

Figure 23. Flow chart of mineral formation from a melt, called the Bowen's reaction series. For granitic rock, temperatures may range from about 1200 to about 650 degrees Celsius during the period of crystallization. Not all minerals may be present depending on the starting composition of the magma.

Pegmatite and aplite dikes are planar to lens-shaped intrusions, generally lighter in color than the surrounding host rock, that are made from the last melt of a pluton to crystallize. As a pluton crystallizes, the remaining melt becomes enriched in H_2O and commonly rarer elements such as lithium and boron. The last stage in the pluton's crystallization history is for the water-rich melt to intrude the surrounding earlier formed rocks, including material of the just formed pluton. A pegmatite is produced when a super heated fluid of predominantly SiO_2, Al, K, and H_2O accompanied with the leftover rare elements of pluton, is injected into joints or fractures and then crystallizes, resulting in spectacularly large euhedral (well-formed) crystals. Some crystals in pegmatites can be as large as 3 m in length. Commonly pegmatites have concentrated amounts of rare elements, and may form minerals used for gemstones such as tourmaline, aquamarine (beryl), garnet, and topaz. For example, the gemstones mined from near Pala, San Diego County, California have the best quality pegmatite minerals produced in North America. Unfortunately, most pegmatites in the Sierra Nevada are non-gem bearing. Aplite dikes are similar to pegmatites in their source of the dike material, being derived from water enriched melt, but aplite cools faster, suppressing large crystals growth. Most aplite dikes are made of fine- to medium-grained, equigranular, potassium feldspar and quartz crystals (Fig. 24). Dikes develop under the same stress configuration as joints, and additionally the magma that injected into the crack assists in expanding the fracture. Aplite dikes tend to have sharp and planar contacts with the host wall rock whereas pegmatite material in the Sierra Nevada commonly occupies lens-shaped to irregular masses in sharp contact with the wall rock. Pegmatite and aplite can also grade into one another within a single dike; commonly the pegmatitic material forms as pods or pockets within the aplite.

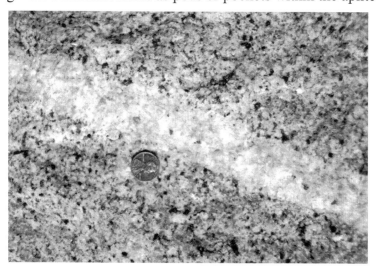

Figure 24. Photograph of aplite dike in the Half Dome Granodiorite at Lake Tenaya. Penny for scale.

GRANITIC TEXTURES AND PROCESS OF CRYSTALLIZATION

The descriptive words used to classify the texture of a granitic rock, the sum of crystal shapes and associations, are a comment on both the evolution of magma and the final solidification state of the rock. The granitic texture represents a growth history that developed in both time and space. Different minerals crystallize at varying times, rates, and abundance according to the composition of the magma, thermal history, and changing pressure. Magma chamber composition is a variable; it fluctuates with injections of new melt and is self-modifying as minerals precipitate out leaving behind a melt reduced or depleted of the chemical components used to form the mineral. The initial points of formation or precipitation of any particular mineral are called nucleation sites. The number of nucleation sites, which is influenced by the amount of water in the melt and the cooling rate, affects the final rock texture. Rapid cooling promotes numerous nucleation sites and simultaneous growth of small crystals; the size of the crystals is limited by competition for space with neighboring crystals. Slow cooling and suppression of the nucleation rate (the number of nucleation sites added in a volume per a unit of time) promotes the growth of large euhedral minerals or phenocrysts. The first minerals to form, generally following those of Bowen's reaction series, are suspended in the melt.

Note that molten rock is a complex material composed of numerous different chemical components that form molecular chains, called polymers, which increase the viscosity of the liquid. The viscosity, the resistance of a liquid to flow, is dependent on the percentage of SiO_2, or in other words quartz, in the melt. For example, basalt, a SiO_2 poor volcanic rock, erupts into lava flows that are relatively thin and spread out over large areas. On the other hand, rhyolite, a SiO_2 rich volcanic rock, forms domes and thick, but short traveled, lava flows. In granitic melts, early-formed crystals may not so much be floating due to buoyancy, instead they may be stuck into a thick paste of molten rock. In contrast, mafic magmas, such as gabbro, commonly have collections of early-formed minerals composed of olivine and pyroxene that sink to the magma chamber floor. Gravity settled crystals form what is called a cumulate texture, which is much like piled sand where the grains are supporting one another. The density of olivine and pyroxene is high, and the SiO_2 content of gabbro is low, giving it a decreased viscosity. Therefore, the weight of the early-formed crystals exceeds the surrounding liquid's resistance to flow, and so may descend through the fluid, although perhaps very slowly. The situation is further complicated because the viscosity of magma evolves depending on the temperature, water content, and percentage of suspended crystals present in the melt.

As crystallization proceeds there is less available room for the newly formed minerals, and once the minerals establish contact with one another they develop subhedral grain boundaries. The last melt filling the interstitial voids in a framework of stacked crystals will solidify in these irregular openings with anhedral forms. The cross cutting relationships between mineral grains, crystal growth forms, size distributions of grains, and mineral assemblage present in granitic rocks are best suited for examination using a microscope (Fig. 25). Nonetheless, the rock texture at the outcrop can be observed to vary between different granitic masses, and if greater attention is applied, variation within a single intrusion may be noted. In summary, the texture of a granitic rock is a sum of a long chain of events that may be deciphered by examining the associations between the constituent minerals, their form, alignment, and nature of contacts between the grains.

Figure 25. Example of granitic rock texture observed in a petrographic microscope; polarized field of view is 2.5 millimeters across. The diamond-shaped crystal at the center is hornblende, irregular gray grains are quartz, and gray grains with black stripes are plagioclase.

THE SIERRA NEVADA BATHOLITH

A batholith is the collection of many individual plutons, not necessarily of the same age or composition, into a large belt of igneous rock- also called a magmatic arc (Fig. 13). All the plutons combined along the JMT belong to the Sierra Nevada batholith and comprise only a small percentage of the entire batholith. Contained in the batholith are associated sets of plutons. These related plutons are grouped together as **intrusive suites**. The Sierra Nevada batholith is one of numerous batholiths formed along the western margin of North and South America from the subduction of oceanic crust beneath the continental crust. Large batholiths also run the length of Baja California and the coast of British Columbia. The Andes of South America is modern setting where a batholith is forming at the roots of the volcanic chains. Furthermore, the older Cretaceous Coastal batholith of Peru has a similar scale and geologic features as that of the Sierra Nevada batholith. Both were formed along the same system of subduction bounding the eastern side of the Pacific basin, both were part of the *Ring of Fire*.

All the plutons were amalgamated into the batholith, one by one, replacing a significant volume of pre-existing rock in the crust. Exactly how a pluton is emplaced into the batholith complex has been a topic of speculation amongst geologists for over a century and it is still not well understood. For the development of any one pluton the older rocks must be moved out of the way by some mechanism. Pluton emplacement mechanisms are two main types: **passive** and **nonpassive**. In passive models, no deformation of the wall rock is caused by the intrusion. Whereas in nonpassive models the intrusion actively deforms the surrounding *country rock,* which is the older rock into which an intrusion is emplaced. Passive pluton emplacement mechanisms are of two basic types: **stoping** and **tectonic opening**. Nonpassive mechanisms are **diapirism** and a related process of **forceful emplacement** or **ballooning**. Combinations of these processes are possible, especially where an ascending pluton travels through several different layers of the crust that have varying material properties under deformation. Most if not all plutons probably originate as diapirs, and then may undergo transformed modes of ascent as based on the contrasts between the viscosity of the magma and that of the surrounding country rock. Most of the plutons were solidified at a depth at approximately 5 kilometers, as based on cooling models of K-Ar radiometric data from hornblende and biotite sampled at various elevations (Kistler and Peterman, 1973). This is in general agreement with depths of up to 4 km as measured by geobarometry of hornblende (Ague and Brimhall, 1988).

Stoping

In the stoping model, the ascending pluton passes through the country rock by excavating blocks of the magma chamber ceiling, which then peel off and sink through the magma (Fig. 26). A major problem in the stoping model is that the pluton roofs are generally smooth, or in other words, do not show irregular blocks calving into the magma chamber. Likewise, large blocks of country rock are generally not found within the plutons. Finally, chemically the composition of the plutons do not suggest melting of the stoped blocks. If a pluton ascends by stoping, the interaction of the magma and the colder wall rock may chill the magma, like ice cubes dropped in a glass of water, and this may arrest the upward motion by promoting crystallization and reducing magma viscosity. This may be bypassed if the size of the stoped blocks are large. Subsiding mega-blocks may explain the relative lack of included wall rock in the plutons and solves some the thermal problems. On the other hand, this mechanism moves the space-generating problem to a greater depth because room must be made beneath the descending mega-block.

Tectonic opening

Tectonic emplacement of plutons injected into the crust generates the space in the pre-existing rock by movement along a fault or a series of faults (Fig. 27). The fault motions are related to the bigger pattern of moving tectonic plates. Along the John Muir Trail is a fault, named the Rosy-Finch shear zone, which probably played a major role in controlling the shapes of the plutons in the John Muir Intrusive Suite (Tikoff and Teyssier, 1992; Tikoff and others, 1999). In the tectonic opening model, the wall rock may undergo

Stoping

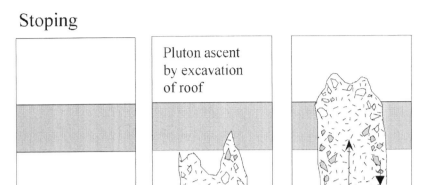

Pluton ascent by excavation of roof

Mined blocks descend along the margins

Figure 26. Diagram showing upward tunneling of a magma body by excavation of roof material and sinking of the stoped blocks. The country rock is not deformed and the magma chamber roof may be irregular. The wall-rock fabrics are sharply truncated and should bear no resemblance to the internal structure of the pluton.

deformation, but it is of a fundamentally different type than that caused by the pluton. In fact, plutons may record deformation as a result of the surrounding tectonic motions. Therefore, in the tectonic opening model, fabrics within the pluton should have a similar pattern to those observed in the surrounding wall rock. A magma chamber, however, is a dynamic feature and may destroy the earlier textures related to opening, and shearing of the surrounding wall rock may not be efficiently transmitted to the viscous fluid. Some examples of pluton emplacement accompanying strike-slip faulting include the Campanario-La Haba pluton in Spain (Alonso Olazabal and others, 1999), and the Las Tazas complex in northern Chile (Wilson and Grocott, 1999). Another mode of tectonic opening may be from incremental dilations leading to the formation of a large pluton by numerous intrusions of dikes (McNulty and others, 1996; Johnson and others, 2001; Glazner and others, 2002).

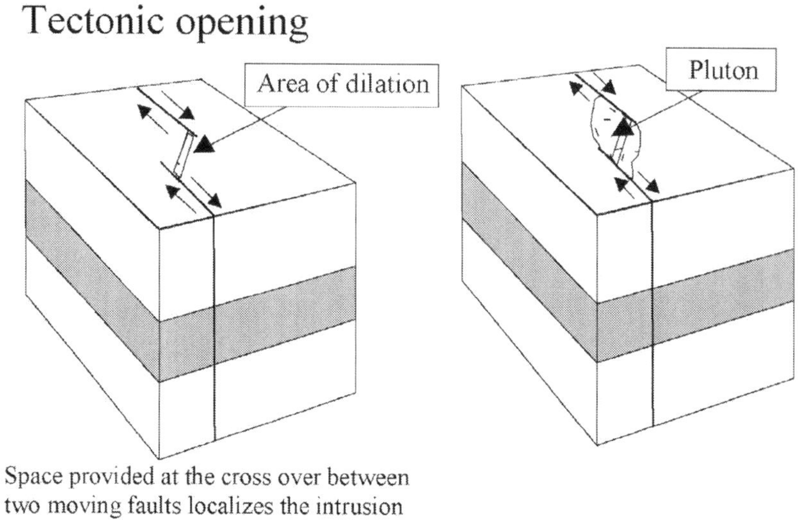

Tectonic opening

[Diagram with labels: "Area of dilation", "Pluton"]

Space provided at the cross over between two moving faults localizes the intrusion

Figure 27. Diagram showing one example of tectonic opening to create the space for a rising pluton. The wall rock may record deformation from faulting, except these fabrics may not be directly caused by the injection of the pluton.

Diapirism

The main type of nonpassive mechanism of pluton emplacement is **diapirism** (Fig. 28). Diapirs are buoyant masses rising due to density contrasts. Fundamentally, this is the main driving force for all plutons. So when the emplacement style of a pluton is discussed, it is really focused on the last increment or time at which the pluton stabilizes its position in the crust. Diapiric rise is a viscous behavior where material is flowing, including the surrounding country rock (Ramberg, 1972; Dixon, 1975; Marsh, 1982; Cruden, 1988; Schmeling and others, 1988; Weinberg and Podladchikov, 1994). For example, the analogy of blobs moving in a lava lamp is appropriate for deeper levels of magma transport. Likewise, density and viscosity contasts between salt deposits and their overburden of sedimentary rocks also result in large diapirs known as salt domes (Jackson and Talbot, 1986). Both stoping and diapirism involve vertical material transfer or exchange. So the difference between stoping and diapirism in the last phase of pluton development is the wall rock response to injection of the pluton. Recently, Tobisch and others (2000) suggested deformation in the Ritter Range pendant is related to the intrusion of the adjacent pluton, requiring a nonpassive emplacement mechanism. Although, fabrics in the deformed metamorphosed volcanic rocks are also in accord with the regional deformation and may better represent tectonic opening. In the diapiric model, the last portion of the pluton can include **forceful emplacement** or ballooning. When the top of the diapir reaches buoyant equilibrium in the country rock, the remaining ascending mass of the diapir injects into the magma chamber, inflating the pluton. Most of the ballooning is accomplished by lateral expansion. This is where a space problem is generated. Either the surrounding wall rock is compressed and deformed to accommodate the expanding pluton, or the wall rock moves away as rigid blocks. A case may be made for forceful emplacement for part of the Mono Creek Granite, where it apparently displaces metamorphic rocks eastward (Bateman, 1992; Tikoff and others, 1999).

Probably the three most studied groups of plutons in the western United States are those crossed by the John Muir Trail. All the above described pluton emplacement mechanisms have been proposed to explain the origin of Sierra Nevada plutons. An important

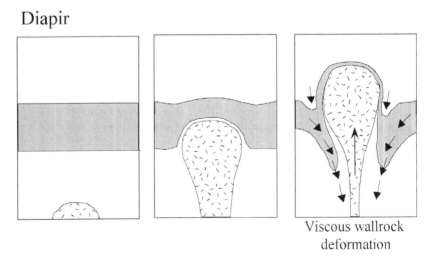

Figure 28. Diagram of the diapir model of pluton emplacement through viscous material. Pluton ascent is accompanied by a combination of wall-rock deformation and return flow. Note the rounded geometry of the magma chamber roof, and that the wall rock should develop foliation concordant to internal fabric of the pluton.

control on which of the above intrusion styles are operative is the nature of the country rock. Temperature and pressure impart a major control on the style of deformation of rock. Stoping is more likely to occur in cooler rocks, and diapirism is expected in hotter or deeper parts of the crust. Of late, geologists have been using mixed-mode mechanism, or combinations of models, to explain pluton development (Paterson and Vernon, 1995). Most importantly, a batholith is a composite feature and develops by a variety of site-specific events.

PRE-BATHOLITHIC ROCKS

Preserved amongst the plutons of the Sierra Nevada batholith are remnants of older rocks. Some of the remnants overlie the plutons and are called **roof pendants**, and were probably the literal roof to a large magma chamber (Fig. 20). Most Sierra Nevada roof pendants actually have steeply dipping contacts- placing them in a position alongside the plutons, and should be referred to as pendants. Thinner elongate bodies of country rock squeezed between two plutons are called **screens or septa** (Fig. 20).

The pendants of the Sierra Nevada are composed of two types of country rock. One variety of pendant is composed of metamorphosed volcanic rocks, which are the volcanic pile that once overlaid the earlier plutons in the Sierra. The other type of country rock is composed of metamorphosed Paleozoic sedimentary rocks that once covered the region before the magmatism in the Sierra Nevada. The metamorphosed Paleozoic sedimentary rocks in the pendants are remnants of the continuous marine deposited strata, rock layering that formed across Nevada and down to the Mojave Desert region. These pendants and screens are one of the more difficult geologic jigsaw puzzles to work on because about 70% of the pieces are missing and these remaining pendants have undergone varying events of metamorphism and deformation. Because most of the roof pendants are deformed, the steeply dipping beds and foliations were mistaken as evidence of nonpassive pluton emplacement. Recently, geologists have been curious about how the metamorphosed rocks in the roof pendants became tilted (Saleeby and others, 1990, Tobisch and others, 2000). Tobisch and others (2000) proposed a model of reverse faults repeating the volcanic section and causing most of the rotation. Moreover, they also believe some amount of tilting is related to deformation caused by the intrusion of the nearby large granitic plutons. A modern analogy in the Peruvian Andes gives clear evidence of volcanic rocks locally being folded to near vertical dips without a substantial amount of fault duplexing (repeated stacked stratigraphic sections) or pluton intrusion.

UPLIFT OF THE SIERRA NEVADA

The uplift of the Sierra Nevada was accommodated by block rotation of the entire mountain belt in response to movement along normal faults on the eastern side of the range (Fig. 29). This block rotation is similar to tilting a table by lifting one side- the Sierras are higher on the eastern side and to the west have gently sloping ridgelines. LeConte (1886) gave an early interpretation of the Tertiary river channels being tilted towards the west. Lindgren (1911) used the straight profiles of the old gold-bearing river channels as evidence of the northern Sierra Nevada block being uniformly tilted, lacking internal deformation such as flexure or disruption by minor faults. Huber (1981a) made indirect estimates on the amount of uplift as based on the degree of river incision. Recently, these ideas are upheld by Dixon and others (2000) who used GPS measurements of the modern tectonic motions of California to show the Sierra Nevada is being transported to the northwest as a single rigid block of crust.

The system of large normal faults that drive uplift of the Sierra Nevada block is a continuation of the faulting style of the Basin and Range structural province to the east in Nevada. This Sierran fault system consists of many interlinked left-stepping fault segments, and defines the western boundary of the Basin and Range Province. The fault segment between Olancha and Bishop is relatively linear, and then northward the faults become more disrupted, or linked to a lesser degree (Fig. 30). The stepped range front is probably at least partially caused by the interaction the horst and grabens of the Great Basin intersecting with the Sierra Nevada block, and by the process of fault growth as individual segments lengthen and eventually connect with one another. Several important aspects of the Sierra Nevada uplift are addressed next, including evidence for modern activity, the along strike amount of displacement and estimate of fault slip rate, and timing of mountain uplift.

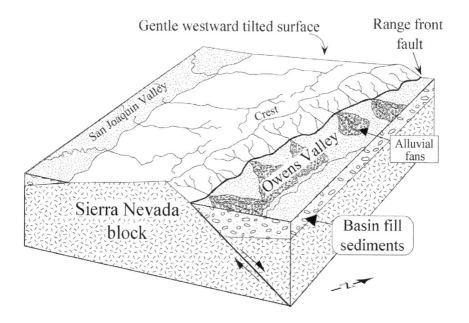

Figure 29. Diagram of the westward tilting of the Sierra Nevada block from normal fault motion along the eastern range front. The Owens Valley side is very simplified in this diagram. The frontal faults are more complex, and Owens Valley has faults running along its east side to bound the next range of the Inyo-White Mountains.

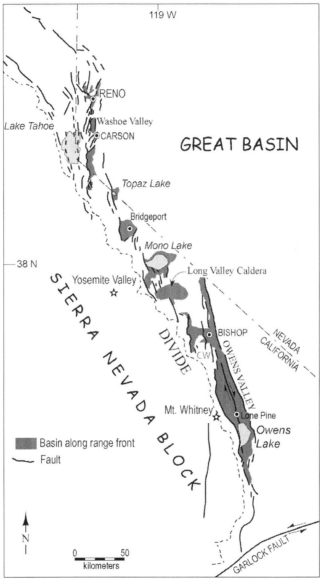

Figure 30. Generalized map of the Sierra Nevada normal fault system and discontinuous basins (faults and basins modified after Pakiser and others, 1964; Bursik and Sieh, 1989; Ramelli and others, 1999; Berry, 2000; Gardner and others, 2000; Henry and Perkins, 2001; Wakabayashi and Sawyer, 2001). Major steps in the crest of the Sierra Nevada follow large left-directed jogs in the range front faults. Regions between the isolated basins have basement rock exposed. CW is the Coyote Warp, a broad flexure that links the Owens Valley fault segment to the range front at Wheeler Crest.

Evidence of Modern Activity

The east escarpment of the Sierra Nevada shows numerous characteristics of active faulting on the basis on landforms. Steep hourglass-shaped canyons separated by triangular-faceted ridges attest to active normal faulting (Knopf, 1918). Fault scarps cut most of the glacial moraines that extend beyond the range front. The large system of alluvial fans is another product of youthful relief. Earthquakes are concentrated east of the topographic divide, making a wide zone along the east escarpment (Fig. 31). This seismic zone contains one of California's greatest earthquakes, which happened in 1872 near Lone Pine (Lee and others, 2001). A study by Lubetkin and Clark (1988) on the fault scarps produced by this historic earthquake documented both normal and right-lateral displacement, demonstrating that modern fault motions in Owens Valley has a tectonic component from the San Andreas fault. However, the Owens Valley fault is not representative of slip along the east escarpment because surface rupture of the Sierra Nevada normal fault which generally has slip lineations oriented down the fault surface.

Commonly the glacial moraines along the eastern base of the Sierra Nevada have been cut by fault scarps. The most recent earthquake to rupture to the surface, was magnitude 6.4, and happened in 1980 on the Hilton Creek fault to the northwest of Bishop. Slip rates on the Sierra frontal faults range from about 0.4 to 1.3 mm/yr (Berry, 1997), and if 1-2 meters of offset are generated per earthquake, the recurrence of earthquakes along any one particular fault segment may be on the order of 800 to 5,000 years. Note that this slip rate is much lower than that of the San Andreas, which is about 3.2 cm/yr. Likewise, motions of major plates at spreading centers or subduction zones is commonly about 3 to 6 cm/yr.

That a major earthquake has not happened on many of the frontal faults, excluding the predominately strike-slip Lone Pine fault, marks the region as one of high risk for earthquake damage, especially at the populated areas of Reno and Carson City. A major earthquake may trigger rock avalanches, or possibly disrupt the Los Angeles aqueduct. The area is relatively sparsely populated, and represents at least a magnitude of order less hazard than the major metropolitans along the San Andreas fault such as San Francisco or Los Angeles.

The distribution of the modern earthquakes measured by

seismograms help define the areas of concentrated deformation, but provides only a partial record of strain because areas of the crust are also slowly deforming without releasing earthquakes. The earthquake data in Figure 31 represents only a very small percentage of events, and if a longer sampling period on the order of 10,000 years became available most of the faults would be clearly delimited by the distribution of earthquakes. During the 1973 to 2007 period a total of 6,576 earthquakes were measured in the map area of Figure 31, and had an average magnitude of 3.5. Most of these earthquakes are in the lower magnitudes (Fig. 32), however the amount of energy released by the larger magnitude earthquakes is significantly greater than that of the combined smaller earthquakes.

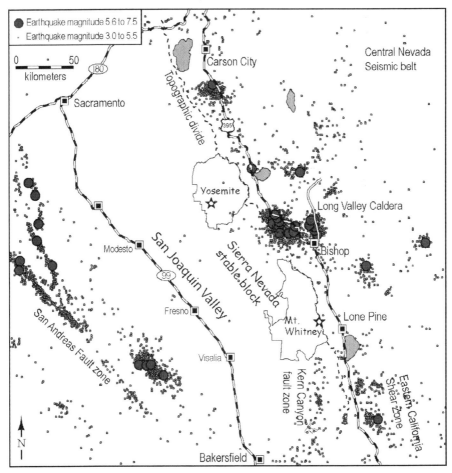

Figure 31. Map of seismic activity in eastern California and western Nevada for the 1973 to 2001 period (earthquake data from PDE

catalog available from the U.S.G.S.). Most of earthquake magnitudes are less than 5. Low recurrence intervals for slip along the Sierran frontal faults, on the order of 800 to 5,000 years repeat time, make large events extremely under sampled in this data set. Area of dense earthquake activity near the Long Valley Caldera is caused by deformation related to the movement of magma, slip events on the Hilton Creek Fault, and motion on faults that lie within the Mount Morrison pendant area. Cluster of aftershocks south of Carson City is from the 1994 Gardnerville 6.4 magnitude earthquake.

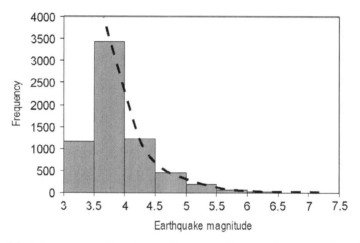

Figure 32. Histogram of earthquake magnitudes for the dataset used in Figure 31. Note the decay pattern of data following the dashed heavy line. Lower magnitude earthquakes are under sampled because of seismogram sensitivity and attenuation of ground motion away from the earthquake hypocenters. Most of the earthquakes are from between 0 to 44-km depth below surface and the mean depth is 6 km. Likewise, seismogram data does not include historic earthquakes, such as the historic Owens Valley earthquake in 1872.

Displacement along the Sierra Nevada Normal Fault

The eastern frontal faults delimit a series of jogs instead of being completely linear, as originally noted by Russell (1887) and Fairbanks (1898). In general, the amount of displacement across a fault tapers or diminishes towards the fault ends. In other words, the area of greatest displacement is commonly at the center of the fault. This geometric pattern can be applied to the elevations of the Sierra Nevada, noting the attenuation of the crest to the north and south of

Mount Whitney area (Fig. 33, Table 2). Total relief between the topographic divide and the alluvial plain also attenuates in a similar pattern.

The pattern of elevation change will be complicated by the summation of displacements for each fault segment of the system, but overall the vertical throw has to be more significant at Mount Whitney because of the greater elevation. **Throw** is the vertical component of displacement for a dipping fault, and it is always less than the total fault displacement. The vertical relief of Mount Whitney above the Owens Valley is 3,318 m (10,900'), and when considering the depth of basin deposits in Owens Valley, the total amount of vertical offset of the Eocene paleosurface (the past land surface) probably exceeds 5,200 m (17,061'). A more detailed discussion of the elevated erosion surfaces is given at the beginning of the Whitney to Forester Pass segment of the guide. The above offset estimate assumes the buried basement-sediment contact in the basin is part of the Eocene erosion surface that once covered much of the Sierra Nevada block. The thickness of basin fill sedimentary and volcanic rocks, as based on a geophysical study of gravity and limited reflection seismicity by Pakiser and others (1964), are approximately 5,486 ± 1,523 m (18,000' ± 5,000') at Mono basin, 2,438 ± 610 m (8,000' ± 2,000') near Bishop, and >1,829 m (>6,000') at Owens Lake. Subsequent studies at Mono Basin indicate basin depths of 1.5 to 2 km (Christensen and others, 1969; Pakiser, 1976). Comparison of these basin depths to elevations of the divide yields a similar amount of minimum offset (throw) as for the Mount Whitney area. Farther north, basin depths at Truckee Meadows of Reno and Eagle Valley at Carson City have maximums at 1.2 and 0.5 km, respectively (Abbott and Louie, 2000). In general, the longer the fault the more displacement can be accommodated by a particular fault. Examination of Figure 30 and comparison of range height and basin depths listed in Table 2 quickly shows that the longer basins are positioned adjacent to more continuous fault segments and the amount of fault offset is greater as reflected by the total relief. Because of the complex segmentation of the range front faults no single value of offset can be stated as representing the uplift of the Sierra Nevada.

Another way of calculating the total offset of the Sierra Nevada normal faults is to use an estimation of the fault slip rate. Near the center of the fault system, the modern range of known slip rates is from 0.4 to 1.3 mm/yr. Projecting these upper and lower values for the past 5 million years gives a rough estimate of offset at about 2,000 to 6,500 meters. Of course, this assumes a constant slip rate through time, which a big assumption indeed. Nonetheless, note that the higher estimate derived from the slip rate data is similar in magnitude as that comparing divide height to basement depth. Note that some of the deeper west-draining canyons, such as the Kings River, have about 1,900 m of local relief between the canyon floor and the divides. If all of this relief was made during the last 5 million years it would equate to a river incision rate of about 0.38 mm/yr, which is a value similar to the above lower fault slip rate. These estimates are also highly dependent on the timing of initial mountain uplift, a point discussed in greater detail later in this section. Even so, it is interesting to see that these rough estimates on fault slip rate, amount of apparent offset along the range front, and rate of canyon incision are close agreement.

The offset along the range front faults provide no inferences on how much of the difference is from the Sierra Nevada block moving up and the Owens Valley floor dropping down, though clearly Mount Whitney stands above everything else in the state, and therefore a certain amount of uplift is required. Note that the higher portion of the range is paired with the most continuous or longest-range front basin at Owens Valley. Finally, wherever normal faults form, the near surface crust deforms as an elastic material that distributes greater uplift and subsidence near the intersection of the fault with the surface. That the Sierra Nevada frontal faults are deep penetrating is shown by the abundant basaltic rocks that have erupted along the structure, such as at Crater Mountain south of Big Pine. Under such conditions this system of major normal faults is even more liberated to accommodate footwall uplift, perhaps releasing the crust to reach complete isostatic equilibrium. Explanations of the Sierra Nevada block as being a completely static region of inherited high topography that was broken to down drop just the basin areas are mechanically improbable. Both the foot wall and hanging wall have their respective motions for which the absolute values do not yet have an original datum for comparison.

Offset markers, also called **piercing points**, that are located on either side of a fault, have never been identified for the Sierra Nevada frontal faults. Estimates on fault offset using the height of the modern divide and the depth to basement rock are minima because of erosion of the footwall rocks. Both range front relief and basin depth in the Bridgeport-Mono-Long Valley Caldera fault segments are non-representative of the overall range uplift pattern because of superimposed local subsidence related to volcanic calderas. Furthermore, higher heat flow in the region may impart an overall broad topographic uplift due to thermal buoyancy (warmer rock is less dense and therefore rises due to isotasy). Wakabayashi and Sawyer (2001) estimated a maximum of about 1-kilometer of late Cenozoic stream incision for the major west-draining canyons of the Sierra Nevada. This represents the minimum amount of uplift of the Sierra Nevada block because the erosion of river channels need not be in equilibrium or keep pace with uplift. Even so, the Palisade-Whitney segment it appears to have about 1-2 km of footwall uplift and the remaining 3-4 km of the total offset is accounted by basin subsidence.

TABLE 2
GENERAL ESTIMATES OF RELIEF ALONG THE EAST ESCARPMENT*

Location	Divide	Scarp	Relief	Depth!	Min.vert.offset
Truckee Meadows	3,047	1,524	1,524	1,200	2,724
Washoe Valley	2,152	1,560	592		
Eagle Valley	2,755	1,524	1,231	500	1,731
Genoa	2,713	1,509	1,204		
Minden Valley-Jobs Peak	3,000	1,500	1,500		
Mono Lake-Log Cabin	3,100	2,000	1,100	2,000	3,100
McGee Mtn.-Crowley	3,313	2,300	1,013	caldera	
Palisade Crest	4,200	2,000	2,200	2,450	4,650
Mt. Williamson	4,100	1,900	2,200		
Mt. Whitney	4,300	2,000	2,300	>1,830	4,230
Olancha	3,600	1,300	2,300		
Coso Junction	2,500	1,300	1,200		
Indian Wells-Owens Peak	2,200	1,000	1,200		

* All measurements in meters
! Depth below surface

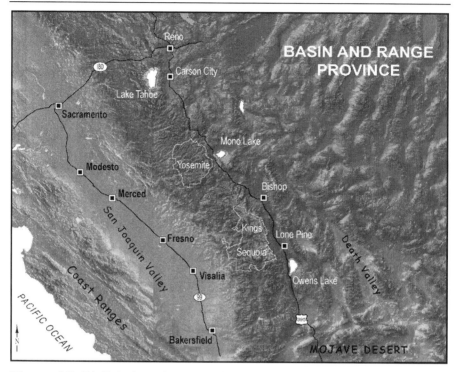

Figure 33. Digital elevation model (DEM) of the Sierra Nevada block and surrounding region showing highs in browns and lows in light blue. The Sierra Nevada block comprises an overall westward sloping surface, in contrast to the highly disrupted, alternating NS-elongate highs and lows of the Basin and Range Province. DEM data is made from global SRTM 90-m resolution survey.

Timing of Uplift

Geologists have speculated or tried to determine when the Sierra Nevada range first started being uplifted as a topographically high feature. Most studies used the development of erosion surfaces that are overlain by Cenozoic volcanic rocks, which were either radiometrically dated or constrained by fossil ages (Dalrymple, 1963, Bateman and Wahrhaftig, 1966, Christensen, 1966). The Sierra Nevada presently causes an orographic rain shadow forcing most of the Great Basin into being a desert. Axelrod (1957) examined fossils from the Sierra Nevada and state of Nevada region to constrain the initial development of this rain shadow and therefore the timing of mountain uplift. This data was used to suggest most of the Sierra Nevada topography formed after the Pliocene (Axelrod, 1962).

Unfortunately, it is difficult to determine the initial time of uplift from the paleontological evidence because of time resolution of the faunas and difficulty in separating the affects of the Sierra Nevada from the background climate change. Finally, new work suggests overall lowering of elevations across the Great Basin (Wolfe and others, 1997), but isolation of the Sierra Nevada block still forces a rain shadow to the east.

From a variety of geological data, the most substantial phase of uplift began approximately 5 million years ago (Unruh, 1991; Wakabayashi and Sawyer, 2001), and the Sierra Nevada is still actively rising while the adjacent Owens Valley is dropping. In the northern Sierra Nevada, near Reno, Henry and Perkins (2001) dated volcanic rocks from the Verdi-Boca basin that were once deposited continuously between the present Reno basin and the uplands near Truckee. This work documents uplift of the Carson subrange after about 3 Ma, and additionally shows evidence for earlier extensional basin formation at about 12 Ma. These studies are in general agreement with the cooling ages of the minerals apatite and sphene, which provide a direct approach to the timing of uplift in the Sierra Nevada. Work by Dumitru (1990) and House and others (1997) have interesting results, indicating early uplift between about 70 and 60 million years ago, a prolonged period of little change or cooling between 60 and 15 Ma, and then initiation of uplift at about 15 Ma. This information is in accord with marine mudstone deposits in the Paleocene Goler Formation at the north side of the El Paso Mountains just south of Inyokern and east of the Sierra Nevada block (Cox, 1982; Reid and Cox, 1989). The Paleocene time interval was from about 65 to 55 million years ago. A marine incursion at this time prohibits high elevations in this region as some speculate may have been inherited from the earlier volcanism and tectonism of the Cretaceous Sierra Nevada magmatic arc. The samples used in the above cooling studies were taken from Yosemite, Kings Canyon, and Mount Whitney and suggest the rocks presently exposed at the surface were about 2 to 3 km depth in the 30 to 15 Ma period. This is in sharp conflict with Eocene age deposits that overlie some of the basement rock. However, these sample areas are not from known Eocene surface areas.

Compatible with the 15 Ma timing of uplift, Chase and Wallace (1986) suggested that at about 10 Ma the region of the Sierra

Nevada had about 1 km of broad elevation. Conrad (1993) described 13.6 Ma volcanic unit interbedded with gravels along the eastern flank of the Inyo Mountains, suggesting that this range and the adjacent Owens Valley were developed by this time. Bachman (1978) described six degrees of westward tilting in lake beds to the east of Big Pine, noting that the beds are about 2.3 Ma in age, places a minimum constraint on the formation of Owens Valley. When the Sierra Nevada formed a mountain range may have been a two-phased process. Uplift is constrained to be younger than about 15 Ma and older than about 3.5 Ma in the southern Sierras (Dalrymple, 1963), and most likely started between 10 and 5 Ma and then the uplift rate probably accelerated after 5 Ma. From 5 Ma to present uplift may have progressed by intermittent bursts separated by slower intervals of uplift as based on the multiple plateaus in the upper Kern River area (Matthes, 1937: 1960). Alternatively, the latter uplift may have been uniform and the rate of weathering was variable due to climate changes.

In as much as the Sierra Nevada block is rising, the Great Valley to the west is subsiding. The remnants of Eocene deposits scattered across the Sierra Nevada are completely preserved in the subsurface of the Great Valley, along with all other subsequent younger deposits. Earlier workers, such as Becker (1891), attempted to relate the subsidence to the uplifted block by having the eroded sediments transported westwards and deposited to load the region of the Great Valley, producing the subsidence. Sedimentary loading may enhance subsidence; however, this is not the main force involved. Tilting of the Sierra Nevada block requires deeper processes operating on the crust, to stretch it thinner, and to provide space for the down-going part of the tilted block. Presently, motion of the Sierra Nevada block is considered as an integral part of the Basin and Range province where the crust has been pulling apart for various periods in the Tertiary (Best and Christiansen, 1991; Catchings, 1992; Axen and others, 1993). The onset of rapid cooling at about 15 Ma in the Sierra Nevada batholithic rocks is concurrent with the formation of the northern Nevada rift in the state of Nevada (Zoback and others, 1994).

The driving forces causing Basin and Range extension are complex. Earliest phase of extension may be related to gravitational collapse of over-thickened crust following the Sevier orogeny (e.g., Glazner and Bartley, 1984). Under-plating or heating of the crust in part may be related to mantle plumes, such as passage of the Yellowstone hotspot. Later phase extension, and uplift of the Sierra Nevada, may be linked to development of the San Andreas fault (Atwater, 1970). Collectively, the basin and range has superimposed all the above geologic processes so that isolating the relative contribution from each is difficult.

Phase of Earlier Uplift

The story of uplift is a little more complicated than outlined above. Sams and Saleeby (1988) presented evidence that the rocks exposed in the southern tail of the range, near the Mojave Desert, have a greater amount of uplift than to the north. The metamorphic rocks exposed in the south formed under higher pressures than rocks to the north, indicating they came from a deeper level in the crust. These deeper rocks may have been uplifted just before the Eocene period, possibly by gravity acting on the structurally thickened and thermally weakened crust from the main period of Cretaceous magmatism to pull apart or extend the uppermost crust (Wood and Saleeby, 1997). Therefore, the basement rocks of the Sierra Nevada have experienced more than one period of uplift. How much topography from the ancestral Cretaceous magmatic arc and/or latest Cretaceous to Paleocene unroofing of the batholith persisted to the late Cenozoic remains problematic.

The volcanic pile that once covered much of the Sierra Nevada batholith, similar to the volcanic rocks of the Andes of South America, were eroded away to expose their plutonic roots after 80 Ma and before the Eocene time. As discussed above, the earlier uplift happened between about 70 to 60 Ma (Dumitru, 1990). In the northern Sierra Nevada Eocene river channel deposits lie directly on top of the batholith. These rivers most likely flowed across a landscape similar to the Pacific flank of the Andes, carrying with their transported gravels the gold nuggets from which numerous placer workings were mined throughout the California gold rush. Preserved Eocene channel deposits become less common southward on the range (Christensen, 1966). Little evidence of the Eocene land horizon remains preserved as flat topped mountain summits such as Mount Conness north of Tuolumne and possibly the summit plateau of

Mount Whitney. Additionally, Eocene river channels also overlie the Peninsular Range batholith in southern California.

The ancient relief of the area pre-dating the 5 Ma to present uplift of the Sierra Nevada block has a minimum of 500' and locally approached 2,000' (Lindgren, 1911). To the east, the paleo-elevations probably were significantly higher before about 45 Ma. Folding and thrust faulting built a Late Cretaceous mountain belt known as the Sevier orogeny, that was the precursor to extensional collapse of the Western North American continental crust. Subsequently, thinning of the crust probably reduced the overall elevation of the Great Basin area (additional discussion in Small and Anderson, 1995; Wernicke and others, 1996; Wolfe and others, 1997). In apparent agreement, recent work suggests that the Sierra Nevada were a high feature before 16 Ma up to present (Chamberlain and Poage, 2000; Poage and Chamberlain, 2002), however, this data set uses isotope values from clay minerals sampled in the lee side of the range, mostly in Nevada, and therefore may also be influenced by the uplift history of the Inyo-White Mountains. Finally, these recent studies do not consider the Paleocene marine deposits in the northern Mojave Desert (Reid and Cox, 1989). This final point opens the broader question about the location of the Eocene coast line.

Tilting of the Sierra Nevada block resulted from a complex history of different phases of deformation apparently spread across the Cenozoic. Ancestral topography may be inherited from the initial uplift and unroofing of the Sierra Nevada batholith in the latest Cretaceous. Early Cenozoic extension disrupted the continental margin, probably reducing the overall elevation. Several phases of late Cenozoic extension certainly is mechanically linked to fault movement along the range front. Geomorphic erosion surfaces of the Sierra Nevada suggest varying periods of erosion and stability. How high the early Sierra Nevada were in throughout the Cenozoic remains an active field of research and much additional work is needed before a complete topographic evolution is determined.

GLACIAL HISTORY OF THE SIERRA NEVADA

Introduction

The recent uplift of the Sierra Nevada freshly exposed the vast white granitic rocks of the batholith and the less abundant rusty-red weathering slivers of metamorphic rocks. This event alone would make the Sierra Nevada high, but crude and unfinished. Over the last two million years, numerous glacial events put the aesthetic touches on the Sierra Nevada. Ice sculpted the mountains to a work of art that painters, photographers, and writers try to capture today. The John Muir Trail caresses the carved out landforms of glacial valleys, aretes, cirques, and tarns, yet this trail also traverses the shavings and left over chips of the mountains preserved, at least temporarily, in glacial deposits such as lateral and terminal moraines, and boulder erratics. These leftovers are the clues geologist use to estimate how many times and when the forces of ice eroded the mountains.

Before the glaciation of the Sierra Nevada the mountain range consisted of an uplifted extensive flat surface, and it was developing an immature landform (eroded) through stream incision. In other words, a relatively flat area or region of low rolling hills was tilted upward and subsequently had V-shaped stream valleys cut into the land surface. The following glaciation events presumably emphasized the early drainage pattern by carving deeper down into the valleys and modifying them to a U-shaped cross section. However, stream and glacier erosion has alternated through time. The lower part of the canyons were probably only eroded by streams.

Glaciers form where the average temperature permits net accumulation of ice, which is generally defined as the region above an elevation or point called the **equilibrium line**. At this point, the glacier neither gains nor losses in its ice mass. Below the equilibrium line the glacier undergoes reduction of mass, a region called the **zone of ablation**, by melting, sublimation (direct loss of ice into to the vapor form without crossing the intermediate liquid state), and possibly by calving of ice blocks from the terminus of the glacier. The equilibrium line is not fixed through time, it varies with annual changes in the climate. Consideration of the gains and losses of the entire system comprise the **glacial budget**.

The role of glacial erosion is clearly expressed in the Cordillera Blanca, central Andes of Peru. Here an 8 million year old plutons of the Cordillera Blanca batholith was uplifted by a major normal fault, forming some of the highest elevations in Peru (Petford and Atherton, 1992; McNulty and others, 1998). The Cordillera Blanca has over 30 summits exceeding 6,000 m. Multiple large glacial carved valleys were cut deep into the plutons. Considering the high elevations, glacial erosion was probably more important than stream erosion throughout the formation of the mountains. Because rocks of the Sierra Nevada are old, there is confusion about how much of the topography formed at what times. The example from the Cordillera Blanca shows that significant glacial valleys can be eroded in less than 8 million years. The development of glacial valleys in the Sierra Nevada is compatible with the interpretation of mountain uplift starting about 5 million years ago.

Glacial features

The types of glacial carved features are numerous, and each tell something about a small part of the larger system. A **cirque** is a semi-circular steep wall formed at the head of the glacier by the excavating power of the moving ice (Fig. 34). When the cirques of adjacent glaciers approach and merge into each other, they form sharp-crested divides called **aretes**. A glacier carves into the rock at its base mainly by the process called **plucking**. Melt water seeps into cracks and joints in the rock, freezes, and when the glacier moves it pulls out blocks of rock and carries it along imbedded in the ice. The glacier ice contains various sizes of rock, from house-sized boulders to rock flour (very finely pulverized rock). **Tool marks** are features left upon the valley floor by rocks in the ice being dragged along with the glacier. Striations (thin, straight scratches) and grooves are the more common types of tool marks (Fig. 35). **Glacial polish,** an uniform glass-like smooth surface on the rock, is also a type tool mark, which forms by very fine rock flour being rubbed along the valley floor by the ice. **Percussion or chatter marks** are a series of crescent-shaped fractures in the rock formed in a row by a large rock being dragged across the glacial carved surface (Fig. 36). **Flutes** are exaggerated grooves; some are several feet deep, U-shaped, and straight. **Potholes** are smooth scooped out holes in the rock made by stream water flowing beneath the glacier. The stream moves the stones about in a circular motion that slowly wears down the rock;

commonly potholes may contain the spherical-shaped millstone that developed inside the hole. A classic location for seeing glacial pot holes is in Tuolumne Meadows, along Highway 120 at Pot Hole dome. **Tarns** are depressions in the valley floor made by the glacier and subsequently filled by small lakes after the glacier melted. Most lakes in the Sierra Nevada are not filled by sediment, indicating they are recent landforms. Opposite of tarns, are positive features in the valley floor shaped as asymmetric glacial-carved domes called Roche moutonnée. **Roche moutonnée** are elongate domes that have gentle dipping up-valley slopes and abrupt down-valley cliffs from which the glacier would pluck or excavate rock. Roche is the French word for rock.

All the above-mentioned processes combine to modify a V-shaped valley into a U-shape profile, which is described in greater detail in the Mather to Muir Pass chapter. **Hanging valleys** form where a tributary glacier enters a larger-sized glacier. Hanging valleys are cut off U-shaped tributary canyons that are high on the sidewall of a larger glacial valley (a classic example is at Bridal Veil Falls in Yosemite Valley).

After a glacial event the ice melts away, leaving behind any material it was carrying. Glacial deposits are composed of rock material (boulders, pebbles, and sand) once transported by ice. The main type of deposit is called till. **Till** consists of all the material a glacier was carrying, and then deposited when the ice melted. Till may be deposited into several geometries depending upon the location about the glacier. Till deposits formed along the sides of a glacier create **lateral moraines**. Till deposited at the glacier's terminus are called **terminal moraines** (Fig. 34). If the glacier is retreating and melting, a series of spaced-out terminal moraines along the valley floor are called **recessional moraines** (Fig. 37). Both terminal and recessional moraines may form dams to lakes (e.g., Convict Lake, Sally Keyes Lakes, and Fallen Leaf Lake south of Lake Tahoe). Large boulders left upon the low-angle slabs of glacier polished rock are called **erratics** (Fig. 38). For additional explanation of glacial features and process see the book *Modern and Past Glacial Environments* by John Menzies (2002).

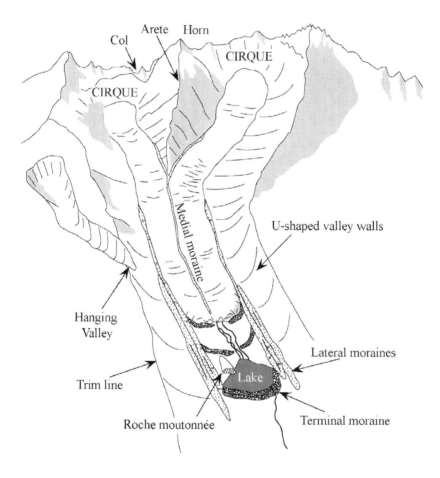

Figure 34. Diagram of the main features in alpine glaciated topography.

Geology of the John Muir trail 79

Figure 35. Glacial striations and polish from a slab in the Tuolumne Meadows region. Swiss army knife for scale.

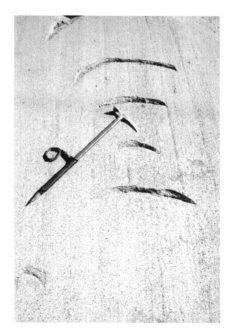

Figure 36. Glacial chatter marks are crescent shape percussion fractures from a rock being drug along the base of the flowing ice. Glacial scoured slab on north flank of Mount Lyell. Ice flowed from the top of the photo towards the bottom parallel to the linear striations and perpendicular to the chatter marks. Ice axe is nearly 3-feet long.

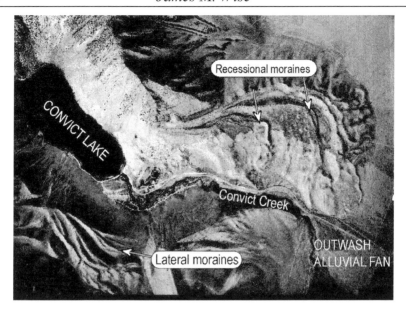

Figure 37. Aerial photograph of recessional and lateral moraines from the Tioga stage at Convict Lake. Convict Lake is about 1.4-km long.

Figure 38. Glacial erratics on top of Lembert Dome, Tuolumne Meadows. View is towards the west.

Glacial events

In 1909, Penck and Brückner identified four glaciations in the European Alps, which they called the Würm, Riss, Mindel, and Günz.

The major continental ice advances in North America, from older to younger, are called the Nebraskan, Kansan, Illinoian, and Wisconsinan. The Pleistocene oceanic sediment record indicates that approximately 17 periods of glaciation, for which evidence of at least five glacial events are present in the Sierra Nevada. Past climatic conditions are extracted from deep sea floor sediments by drilling core sample (Martinson and others, 1987), and provide a better-constrained climatic history (Fig. 39). Periods of glaciation also change the global sea level, and can generally be related oxygen isotope data (e.g., Haq and others, 1987). Multiple periods of cooler climate and glaciation are in part caused by periodic changes in the Earth's orbit (eccentricity) and rotation configuration (obliquity and precession), called Milankovitch cycles. Additional geologic factors may also influence the changes in climate, including major volcanic eruptions, ocean circulation patterns, and uplift of mountain belts.

In the Sierra Nevada, four of glaciation events were designated by Blackwelder (1931) with the following names from youngest to oldest: the Tioga, Tahoe, Sherwin, and McGee stages. The fifth possible glacial event is recognized near Lee Vining, and is named the Mono Basin stage. A sixth stage of questionable importance between the Tahoe and Tioga glaciations is called the Tenaya stage. Younger glaciation events tend to remove the older glacial deposits, therefore few places preserve deposits from all five events. Generally, if the older deposits are to be preserved, the younger glaciation event must cover a smaller area, otherwise it will remove the older deposits. Till from the five glacial stages are mainly composed of granite, making the relative dating process difficult. Different age till may be distinguished by the relative positions to one another, and by the degree of weathering and soil profile development. Also considered are the preservation of the overall moraine shape, such as the morainal crests, and the degree of stream erosion on the flanks of moraines. Specific identifying features for the five of the six stages of glaciation are described in the following paragraphs.

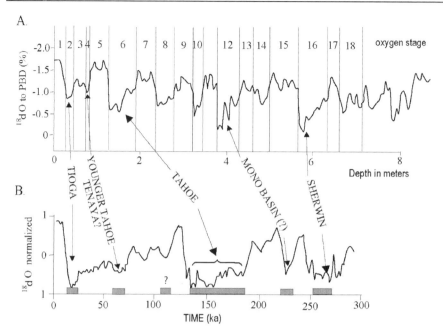

Figure 39. A. Deep sea sediment sample from drill core V28-239, showing multiple periods of cooler climate (Shackleton and Opdyke, 1976). B. Orbitally based chronostratigraphy of stacked oxygen isotopes from Martinson and others (1987), derived from microfossils in deep sea floor sediments. Peaks represent the warmer interglacial periods. The approximate periods of Sierra Nevada glacial stages and other potential periods of glaciation are marked by the gray bars.

McGee Stage

McGee till is the oldest recognized glacial deposits in the Sierra Nevada. The deposits are small and isolated, found on the eastern side of the Sierra Nevada (Blackwelder, 1931; Putnam, 1962), and they do not occur along the JMT. The date of this glacial stage is uncertain. It is younger than 2.6 Ma (Dalrymple, 1963; Rinehart and Ross, 1964), which is the age of a basaltic flows and cinder that underlie the till. The age is from a basalt flow underlying the McGee till at an uplifted surface south of Convict Lake. Schaffer (1997) erroneously interpreted the McGee till at this location as being highly weathered granitic intrusion. The till contains boulders derived from the Round Valley Peak Granodiorite, as originally noted by Rinehart and Ross (1964), and these boulders have inconsistent foliations, indicating that they are not in place. Rinehart and Ross (1964) reported the presence of striated boulders, and found several clasts of metamorphic rock, features that Schaffer (1997) failed to note. Also, the base of the till follows a map pattern intersection with topography that indicates a subhorizontal basal contact, not a steeply dipping contact as would be expected if it were an intrusion. The geometry of the till deposits in map view define a T-shape, not a typical pattern for a pluton. And lastly, the adjacent metamorphic rock and the Late Triassic Wheeler Crest Granite are not intensely weathered, as Schaffer (1997) proposed for the McGee till. Another site of pre-Sherwin till near Rock Creek may belong to the McGee till (Sharp, 1968), but these deposits have yet to be directly dated. At Owens River gorge, glacial outwash fans overlie basalt flows dated at about 3.2 Ma and are in turn overlain by the Bishop tuff (Sharp, 1968). These deposits may record both the Sherwin and McGee stages.

Sherwin Stage

Sherwin till makes the most volumetrically large deposits along the eastern side of the Sierra Nevada. It has been extensively modified by erosion, being dissected by lengthy drainages, and forms rounded hills lacking any of the original moraine geometry. The till contains few boulders rising above the land surface, and when boulders are present they are rotten and have potholes weathered in them. The Sherwin till was overlain by volcanic deposits of the Bishop Tuff (Sharp, 1968), which erupted 760,000 years ago (Bailey, 1989), demonstrating that the Sherwin stage is at least older than this time. Sharp (1968) interpreted the age of glaciation to be about 40,000 years older than the Bishop tuff as based on the degree of weathering. The JMT intersects three possible exposures of Sherwin till. The Sherwin stage of glaciation probably produced the greatest amount of erosion in the Sierra Nevada canyons, including establishing the main forms of Yosemite Valley. The classic east side locations of Sherwin till is north of Mono Lake and south of Crowley Lake. The road cuts along Highway 395 north of Mono Lake, near Mono Pass, provide a particularly well-exposed area for examining the Sherwin till.

Mono Basin Stage

Mono Basin till has not been identified along the JMT, and is only preserved in special situations on the eastern side of the Sierra Nevada. The Mono Basin till occurs in places where it suggests it is older than the Tahoe till (Sharp, 1969). In general, the overall moraine form of the Mono Basin till is still preserved, which excludes it from being deposits of the Sherwin stage. The relative weathering of moraine deposits show little difference between the Mono Basin and Tahoe tills (Burke and Birkeland, 1979). In the Devils Postpile region, the Bishop tuff was mostly eroded away before the eruption of the Basalt of Devils Postpile (Huber and Rinehart, 1967), suggesting major glaciation between the Tahoe and Sherwin stages. The main location for the Mono Basin till is at the mouth of Bloody Canyon, southwest of Lee Vining, where a pair of moraines managed to escape modification by the larger Tahoe stage glaciation. If these moraines do represent a separate glaciation and not just a change in the Tahoe glacier flow direction, the Mono Basin till must then be older than about 160,000 years and younger than 760,000 years. Phillips and others (1990) dated the Mono Basin till at 103 ka, an age younger than the Tahoe till. In their interpretation, the Tahoe till has older and younger moraines separated by the Mono Basin till. Clearly additional work is needed on the Mono till at Bloody Canyon and at other locations.

Tahoe Stage

Tahoe till is composed mainly of weathered granitic boulders that have solid to rotten cores (decomposed granite). Overall, fewer boulders protrude above the moraine surface than for the Tioga till, and the Tahoe morainal crests have been eroded down, although the bulk shape to the moraines may still be recognized. The Tahoe till contains a higher percentage of clay from weathering than does the Tioga till (Burkins and others, 1999). The lateral moraines of the Tahoe stage form thick deposits at the base of many of the eastern Sierra Nevada canyons. Higher up most of these deposits were removed by the Tioga stage glaciers.

Birkeland and others (1980) reported age determinations from basalt flows that are interbedded with the till to constrain its time of formation as older than 62 ka ±13 ka and younger than about 126 ka ± 25 ka. Gillispie (1982) reported two $^{40}Ar/^{39}Ar$ method dates from basalt underlying what was interpreted as Tahoe till, placing limits at about 118 and 131 ka. On the other hand, dating of desert varnish on boulders in the till at another location suggest the deposits were in place before about 143,000 years B.P. (Dorn and others, 1987).

Phillips and others (1990) dated moraines called the older Tahoe till at 133 to 218 ka, giving a mean age of about 207 ka. Diatoms from Owens Lake indicate increased moisture between 115 and 170 ka (Bradbury, 1997). Oxygen isotope data from the same location give a cooler climate between 120 to 154 ka (Menking and others, 1997). In summary, the Tahoe stage glaciation probably occurred around 125,000 to 160,000 years ago. Some of the conflicting younger moraines referred to as Tahoe stage are perhaps best considered as part of the Tenaya stage.

Tenaya Stage

Tenaya till is only locally documented in the Sierra Nevada. It is unclear if these deposits belong to a substage of the Tioga or Tahoe tills. Sharp and Birman (1963) and Birman (1964) presented moraine-weathering data having characteristics intermediate to the younger and older tills. Similarly, lateral moraines mapped by Sharp (1972) in the Bridgeport region are nested between the Tahoe and Tioga stage till, and overall weathering characteristics are more similar to the Tahoe till. Bradbury's (1997) diatom data from Owens Lake gives a period between about 62 to 70 ka that is compatible with glaciation. Phillips and others (1990) distinguished older and younger Tahoe till, with the latter being dated at 60 ka. They dated a moraine at 24 ka and suggested that it belongs to the Tenaya stage. It seems likely that the moraines Phillips and others (1990) attributed to the Tenaya stage belong instead to the Tioga stage glaciation. What they call the younger Tahoe till is perhaps best placed in the Tenaya stage. The Tenaya stage seems to lie in the 60 to 70 ka interval. Deposits from this event have yet to be located along the JMT.

Tioga Stage

Tioga stage deposits represent the most recent glaciation event in the Sierra Nevada and are found throughout the length of the JMT. The moraine crests tend to be well preserved, accompanied with large boulders scattered on top of the moraines. The boulders of granite are not weathered. Where glacial polish is found it almost invariably was formed during the Tioga stage. The Tioga moraines are typically nested inside older moraines of the Tahoe stage when both these age deposits are present. Birkeland and others (1980) give the Tioga stage as being older than about 9,800 years. Whereas, desert varnish ages suggest that the glacial deposits are older than about 13,500 years B.P. (Dorn and others, 1987). Sediments in Owens Lake give the Tioga stage glaciation at about ~13,500 to ~24,500 years B.P. by analysis of clay, carbonate, and diatom content, and crossed referenced with oxygen isotope and radiocarbon age data (Menking and others, 1997; Bradbury, 1997; Bischoff and others, 1997; Benson and others, 1998). Throughout most of the glacial stages melt water from the eastern Sierra Nevada passed through a series of lakes, starting with Owens Lake and continuing through China, Searles, and Panamint Lakes, and then ending at Lake Manly in Death Valley. Similarly, Marine oxygen isotope data indicate cooler climate during the 15 to 25 ka period. Phillips and others (1990) dated the Tioga till at Bloody Canyon at about 21.4 ka, using method that measures the amount of cosmogenic isotopes that develop at the surface of exposed rocks. In summary, the age of the Tioga stage is approximately 14,000 to 25,000 years before present.

Rock Glaciers

There are more than one hundred active rock glaciers in the Sierra Nevada. **Rock glaciers** are composed of blocks of rock bound by interstitial ice (filling the holes between the rocks). They tend to be in north-facing cirques, are typically less than 1 km in length, and make lobate blocky deposits (Fig. 40). The material at the toes of the rock glaciers are loose, because the rocks having just melted out from the ice are not completely stabilized or settled. These deposits are the remnants of the last ice age. In previous mapping, deposits closely related to rock glaciers were called the Recess Peak, Hilgard, and the Matthes tills. These deposits are best called the Recess Peak. Through carbon dating, Clark and Gillespie (1997) showed that they formed before ~11,200 years ago, making them the waning part of the Tioga stage glaciation. Although, Clark and Gillespie (1997) suggested that the Recess Peak till represents a new glacial resurgence or advance.

Historically, there was confusion about identifying the glacial stage of moraine deposits because in the Yosemite region only the first three stages can be recognized and they were described using different terminology developed by Matthes (1930). He recognized three stages he named the Wisconsin, El Portal, and Glacier Point. Blackwelder (1931) believed Matthes' Wisconsin stage included both the Tioga and Tahoe stages recognized along the eastern side of the

Sierra, and the El Portal and Glacier Point to be equivalent to the Sherwin stage. The terminology proposed by Blackwelder is more broadly accepted because it avoids confusion with glacial events elsewhere in the United States. The term Wisconsin is standard in describing the latest glacial advance in the Midwest and northeast U.S. The glaciers in the Sierra Nevada were of the alpine glacier geometry whereas the glaciers elsewhere in the United States formed large ice sheets, thus the preference of terms that has no direct implied correlation between glaciers of two different types. Attempts at directly dating the moraines, or dating materials interbedded inside the moraines, help to clarify the relative timing and number of events (Fig. 41). Controversy still persists in the nomenclature and resolution of these methods.

In this guide the description of glacial deposits along the John Muir Trail uses the classification scheme of Blackwelder (1931), and is the first regional attempt at unification of terminology along the crest of the Sierra Nevada. U.S. Geological Survey 15-minute geologic quadrangle maps use local relative age glacial unit names and have not been standardized. The time correlation of glacial deposits is difficult work and still an active line of research in the Sierra Nevada. The glacial stage designations presented in this guide have drawn from numerous published sources and the author's field observations. These descriptions should be considered preliminary, established from map relations and weathering characteristics, but not including more rigorous dating methods. Standardizing the glacial terminology provides an initial framework to be modified and advance our knowledge of glaciation in the Sierra Nevada.

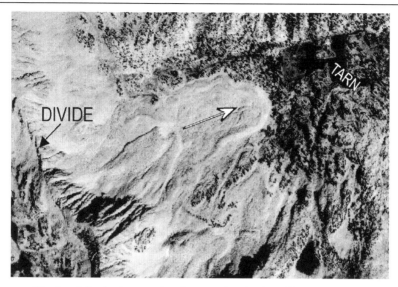

Figure 40. Aerial photograph of a rock glacier (light mottled area at center of picture) in a cirque in a tributary to McGee Creek, south of Mammoth Lakes. The arrow shows movement direction of the rock glacier. Dark spotted areas are stands of pines.

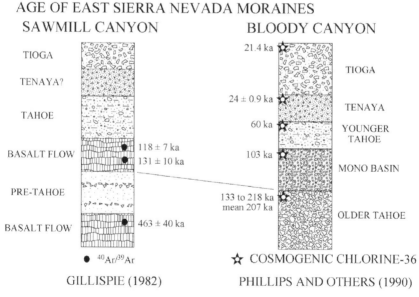

Figure 41. Diagram comparing dated glacial moraines from two locations along the eastern side of the Sierra Nevada.

INTERPRETING GEOLOGIC MAPS

Geologic maps are the fundamental method of recording information about rock types and their distribution. This guide is intended for use with the included simplified geologic maps. The previous mapping of the Sierra Nevada, especially in the regions along the John Muir Trail, were at a scale of 1:62,500. This scale was not appropriate for showing the details of the geology along the JMT, so the geology was re-mapped at a scale of 1:24,000, only including areas adjacent to the trail. This type of map is called a strip map. For clarity, the maps are shown only with the major contour intervals, generally either 100 m or 200 feet.

To be able to read geologic maps a short explanation of the symbols and meaning of the word **contact** is given here. A contact in geology is the interface between two rock types. A contact on a map is represented by a thin line. If the contact is a fault then the line is drawn twice as thick. Contacts for which one can observe on the ground, fully exposed are drawn on the map using an unbroken line. If the contact is buried by what is called **cover**, such as the unconsolidated material of talus or alluvium, but its relative position is known from projecting its last exposure beneath the cover, then the line on the map representing the contact is drawn dotted. If the contact is covered and its position is uncertain, then the line on the map representing the contact is drawn with dots and question marks. Where an exposed contact position is approximate it is drawn using dashes (Fig. 42). For some of the contacts in this guide reference is made to its elevation so a hiker using an altimeter may easily locate the contact as it intersects the trail. Finally, the use of hand held GPS units with the listed UTM coordinates in the guide can assist comparison of field locations to the geologic maps.

All maps in the guide use UTM zone 11 in the 1927 North American Datum (NAD) coordinate system. Maps are shown with the UTM grid vertical, so note that true north lies 1.1 degrees to the east. Magnetic declination is at 15.5 degrees to the east.

The types of symbols used on geologic maps are text abbreviations for the rock types, and symbols displaying information about the orientation of features in the rock. Lithology abbreviations are given a capital letter representing the age of the rock unit and followed by lower case letters giving an abbreviation of the rock unit name. For example, the Half Dome Granodiorite map abbreviation would be Khd; K stands for Cretaceous age, and hd for Half Dome. Several rock unit names were modified from the original published usage to keep a consistent format for the entire trail. For each rock type on the map, a description is given in the text.

Common map units along the John Muir Trail are those for recent talus (Qt), and valley fill related to stream-transported detritus filling lakes or flood plains to form meadows (Qal). Slightly older than these are the glacial deposits of the following map abbreviations: modern rock glacier (Qr), Tioga till (Qg), Tahoe till (Qta), and Sherwin till (Qs). Metamorphosed volcanic rocks commonly have "mv" accompanied with an age prefix. It is important in the field to note the color and texture of rocks up close and in the distance. Next, associate the rock's appearance to exposures in the distance and note its position on the map. Then its abbreviation may be used to look up the rock type description in the guide.

Orientation data for tilted layers in a rock are indicated by a **strike and dip** symbol (Fig. 43). Rock cleavage, joints, and foliation use similar symbols. Other geologic features that can be described for orientation are the alignment of cobbles in a conglomerate, and even the direction of aligned minerals such as hornblende.

In reading a geologic map it helps to locate oneself on the map by examining the shape of the contour lines that represent ridges, slopes, cliffs, and valleys. Next locate the geologic feature to be examined and note its position relative to your position. For example, you may be standing on the JMT and have a nearby peak in view- to determine where both points are on the map, find the symbol representing the rock type composing the peak and check the legend of the map to find out what the symbol means. The position of lakes and streams also provide useful markers in comparing one's location on the JMT or for locating geologic contacts. A compass to orient the maps and take bearings on surrounding features may be useful for those following the guide.

A more advanced map reading technique involves following the trace of a contact as it intersects the topography. From the way the contact deflects as it crosses ridges or valleys, the orientation of the contact can be determined (this technique is referred to as the "rule of V's"). The line of intersection between a vertical plane and topography results in a straight map trace as it crosses ridges and valleys. A horizontal plane intersecting topography follows the shape of the contour lines. For an inclined plane intersecting topography one may determine its orientation from the way it deflects when crossing ridges or valleys.

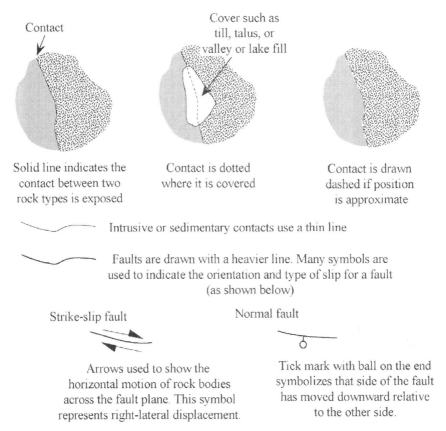

Figure 42. Conventions of symbols for geologic contacts on maps.

EXPLANATION OF STRIKE AND DIP

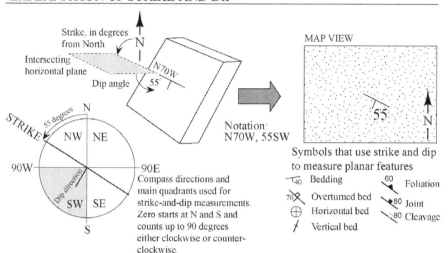

Figure 43. Explanation of the strike and dip symbol. Strike measurements are recorded by two separate systems: the first uses the quadrant method where 360 degrees is broken into 90 intervals in the NW, NE, SE, and SW, and the second uses the full 360 degrees measured from north in a clockwise direction (azimuth method).

OVERVIEW GEOLOGY OF THE SIERRA NEVADA

The Sierra Nevada batholith is a composite of at least three major pulses of magmatism related to subduction of oceanic crust in the Late Triassic, mid-Jurassic, and throughout the Middle to Late Cretaceous. The batholith intruded metamorphic country rock that belongs to both the Paleozoic continental shelf and Mesozoic accreted terranes of the Western Metamorphic belt along the western foothills area (Fig. 44). For most of the Paleozoic, the western margin of the North American plate formed a large continental shelf similar to the offshore region of the modern Atlantic coast. In this area a thick section of marine sedimentary rocks, such as limestone and shale, was deposited and is collectively called a miogeocline or passive margin. In the mid-Paleozoic the continental shelf and slope collided with an assemblage of deeper marine rocks, called the Roberts Mountain Allochthon (allochthon describes a package of rock that was transported). Both the continental and slope strata are still preserved in the roof pendants of the easternmost Sierra Nevada, such as at

Mount Morrison and the Ritter Range pendants. Farther to the west this stratigraphic section is absent, either not continuing to the west, removed by faults, or was destroyed. The omission of these rocks in pendants to the west may be related to a late Paleozoic terrane collision, called the Sonoma orogeny, or were removed by an early Mesozoic regional left-lateral fault.

The Western Metamorphic belt (WMB) is a composite feature containing complexly deformed rocks bounded by steeply dipping faults, and for the most part, represents rocks that were exotic to North America, called terranes, that were transported and then plastered onto the continental margin. Parts of this belt of rock are related to the above-mentioned Sonomia terrane, and the remainder was added to the continental margin during the Mesozoic. The WMB is divided in half by the Melones fault, an important boundary between adjacent terranes. Rocks to the east of this fault are the lower Paleozoic ShooFly complex and structurally lower late Paleozoic to Mesozoic Calaveras complex, which together are know as the Eastern belt (Hannah and Moores, 1986; Hanson and others, 1988; Hanson, 1991). These units were smashed into the continental margin before intrusion of the Late Triassic granites of the Sierra Nevada batholith.

Throughout much of the Mesozoic, the western continent margin of North America ran parallel to the axis of the Sierra Nevada batholith, and was positioned near the Western Metamorphic belt. At about 180 million years ago the North American plate straddled the equator (Engebretson and others, 1985), had a tropical climate, and an immense intermittent inland shallow sea covered from part of the Nevada to beyond the Mississippi River. By the Cretaceous time, the North America continent moved northward about 30 degrees latitude. The Basin and Range province of Nevada had yet to develop, so that overall the Mesozoic plate margin was probably similar to that of the northern Andes in South America. A significant mountain belt formed by folding and faulting, and accompanied with volcanoes, occupied the region of the present Sierra Nevada.

To the west of the Melones fault, a composite of Jurassic volcanic arc rocks and related aprons of sedimentary rocks has collectively been called the Tuolumne River or Foothills terrane (Edelman and Sharp, 1989; Tobisch and others, 1989; Herzig and Sharp, 1992). This package includes the Jurassic oceanic crust of the

Smartville Complex, and the Peñon Blanco, Logtown Ridge, Mariposa, and Jasper Point Formations, which together were accreted to the continent margin in the mid-Jurassic (Bogen, 1985; Day and others, 1985; Beard and Day, 1987). These rocks were rafted along with the subducting oceanic plate that provided part of the melting system to form the mid-Jurassic plutons of the Sierra Nevada batholith. In fact, collision of these volcanic archipelagos perhaps jammed the subduction trench, now represented by the Melones fault, and shut off the magmatism until the subducting slab re-broke outboard near the present coast of California where related melange rocks of this trench comprise the Franciscan complex. Much of the San Joaquin Valley is underlain by oceanic crust, in part explaining the present day low elevations. Once the Cretaceous subduction zone was established, during a period of particularly rapid plate motions, the voluminous Mid to Late Cretaceous intrusions formed systematic pattern of decreasing age towards the east, culminating in the large plutons crossed by the JMT.

The John Muir Trail crosses several large related bodies of Cretaceous plutons interpanneled with pendants and screens composed of metamorphosed volcanic and sedimentary rock. From south to north, the trail begins in the Cretaceous Mount Whitney Intrusive Suite, and then heads north into older plutons separated by screens of metamorphosed Paleozoic sedimentary rocks in the Twin Lakes/Pinchot Pass area. From here it crosses the metamorphosed volcanic rocks in the Goddard pendant, and then continues into the highly elongate John Muir Intrusive Suite, a set of plutons formed slightly before the Mount Whitney Intrusive Suite. In the Devils Postpile National Monument the trail crosses several young volcanic deposits, including basalt lava flows. To the north of this is the Ritter Range pendant, composed of metamorphosed Cretaceous to Jurassic-Triassic tuffs and lavas. The northernmost section of the trail is in the Tuolumne Intrusive Suite that also developed at about the same time as the intrusive suites to the south.

Through all these intrusive rock bodies the right-lateral Sierra Crest shear zone system interacted during the pluton emplacement and deformed older plutons and metamorphic wall rock (Greene and Schweickert, 1995; Tikoff and Saint Blanquat, 1997). This Late Cretaceous fault follows the axis of the Sierra Nevada batholith, and is mostly to the west of the modern topographic crest. The JMT crosses several different segments of the Sierra Crest shear zone system. Other subparallel fault systems probably were active throughout the Mesozoic. In fact, the Cretaceous batholith largely obliterated an older fault that ran the length of the Sierra Nevada, a structure inferred from the age and composition of displaced metamorphosed sedimentary rocks in roof pendants. This older fault is called the Mojave-Snow Lake fault, named for rocks in the Snow Lake pendant to the north of Yosemite that were displaced approximately 450 km northward from their site of origin in the eastern Mojave desert (Lahren and Schweickert, 1990).

In this guide the geology descriptions are focused on the John Muir Trail. However, during the Cretaceous, other plutons were forming both east and west of the present Sierran crest. In fact, some Cretaceous plutons are known as far east as the Nevada-Utah border. Overall, the JMT explores the Cretaceous magmatic belt. Even so, the Sierra Nevada also had major pluton formation events in the Jurassic and Triassic. Also not reviewed in this guide are the metamorphic terranes of the Western Metamorphic belt. Most of the Cretaceous plutons developed after the accretion of these foreign rock belts of the western foothills and represent additional melts derived from renewed subduction.

The Cretaceous magmatism was shut off when the angle of the subducting oceanic crust lowered to become a flat slab being overridden by the North America plate. The passage of oceanic crust beneath the continent in a horizontal configuration results in mantle rock being isolated from the lower crust of the continent, cooling of the override plate, imparting transitory eastward migration of magmatism, and significant deformation called the Laramide orogeny in the region of Utah, Colorado, and Wyoming. In this area large blocks of basement rock were thrust up through their sedimentary cover and the intervening areas between these blocks formed deep basins that were filled by sediments, most notably those belonging to the Eocene Green River Formation from which spectacularly preserved fossil fish are found. Similar foreland basement uplifts are developing today in northern Argentina where the Nazca plate is undergoing flat subduction. At the same time of the Laramide orogeny, the rock that once covered the Sierra Nevada batholith was stripped off, as described previously on the uplift of the Sierra Nevada. This post Cretaceous and pre-Eocene batholith uplift and exposure appears to be a regional tectonic event because the rocks of the Peninsular Range Batholith of southern California, which in the early Cenozoic were located about 500 km southward into Mexico, and even farther south the Coastal Batholith of Peru were also exhumed and covered by Eocene deposits.

At several different intervals in the Cenozoic, the San Joaquin Valley area formed an inland sea. One particular marine bed deposited northeast of Bakersfield, representing the Temblor Sea in the middle Miocene (at about 15 Ma) contains abundant fossils of giant shark teeth (Dupras, 1985). During this period of time was the transfer of part of the North America continental crust to the Pacific plate throughout the growth of the San Andreas fault. Growth of the San Andreas fault halted subduction in its vicinity and therefore also stopped the Miocene calc-alkaline volcanic arc. It is no accident that the southernmost active volcanoes today, at Mount Lassen, are located eastward of where the San Andreas fault joins the subduction trench offshore at Mendocino, northern California. Younger volcanism east of the Sierra Nevada, such as the ca. 4 Ma Coso volcanic field (Duffield and others, 1980), are not products of subduction, instead these magmas were generated by the same high heat flow that accompanies the stretching apart of the Basin and Range province.

Along the entire length of the trail numerous glacial features are beautifully preserved. Some generalization about the glacial morphology, or landforms, along the JMT should be made. First, the valleys to the north side of the main passes along the JMT tend to be NW trending and more deeply incised than the south-facing valleys. The NW orientation of valleys may be from preferential erosion along the overall NW-striking fabric in the rocks (most of the plutons and pendants are NW elongated). Second, the south-facing valleys tend to be broad, devoid of rock glaciers, and retain more of the moraine material left lower in the valleys, suggesting glacial retreat was more rapid than in the north-facing valleys.

Yosemite Valley was deeply incised into the batholith by two large coalescing glaciers. The Sierra Nevada contains numerous similar valleys, such as Hetchy Hetchy, Mono Canyon, and Kings Canyon, upon which John Muir wrote comparisons on their formation. The equation is simple, the greater the watershed focused into one canyon, the more glacier ice will be combined to provide deeper erosion of the bedrock.

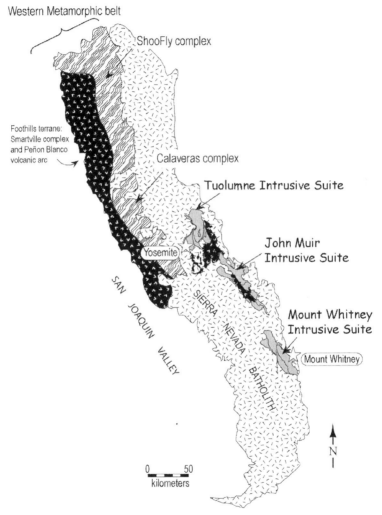

Figure. 44. Generalized geology of the Sierra Nevada. See text for discussion.

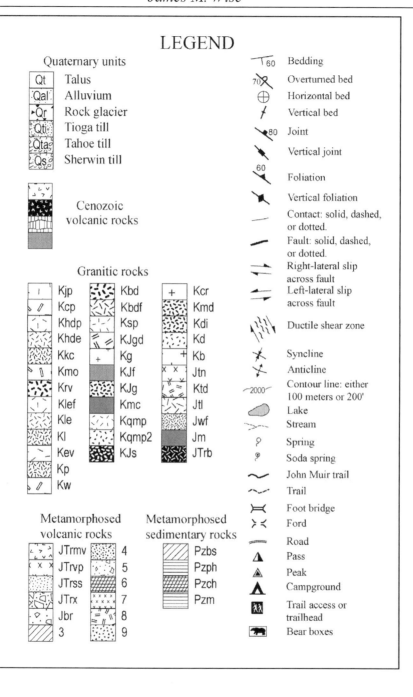

CHAPTER 2

MOUNT WHITNEY TO FORESTER PASS

Access: Whitney Portal, Kern Plateau trails
Distance: 37.3 km (23.2 miles).
Maps: 1-5.

This segment of the guide covers the Mount Whitney Intrusive Suite, and moraine deposits representing at least three separate glaciation events. The scale of the Mount Whitney Intrusive Suite will quickly become apparent once the trail is hiked from Whitney Portal to Forester Pass, a journey requiring the average hiker approximately 2-3 days to traverse.

The John Muir Trail begins at the summit of Mount Whitney (Fig. 45), which is reached by an arduous trek from the Whitney Portal trailhead (located west of Lone Pine, California- see Appendix II). The summit of Mount Whitney (4,416.9 m, 14,491.8'- 1994 survey, waypoint 384470E, 4048690N) is the highest point in the Sierra Nevada and in the conterminous United States. The mountain was named after Josiah D. Whitney, the California State Geologist during the 1860's, by Clarence King in 1864 on the first ascent of Mount Tyndall. Charles Begole, Albert Johnson, and John Lucas made the first documented ascent of Mount Whitney in August of 1873. Near this time, Clarence King had made several attempts but did not bag the true summit. Shortly afterwards in the same year, several other parties climbed to the summit, including John Muir's first ascent of the north couloir, presently called the Mountaineer's route, which is mainly used for descents for rock climbers returning to Iceburg Lake at the base of the east face. The 517-meter (1,700 feet) high east escarpment was first climbed by Robert Underhill, Norman Clyde, Jules Eichorn, and Glen Dawson in August, 1931. About 335 m (1,100 feet) of this spectacular route is technical enough to require ropes.

The summit of Mount Whitney is composed of sills of granite porphyry and aplite injected into the Whitney Granodiorite. The summit plateau is part of an old erosion surface, and west flank of Mount Whitney compose a rubbly mass of blocks and slabs that are loosened from the bedrock by mechanical weathering process of frost heaving. Frost heaving is the tilting, dislodging, and transport of material by the expansion of ice combined with gravity dragging the material down slope. The western side of the mountain makes a gentle-dipping slope, which gradually steepens towards the west. The east face of Mount Whitney forms an impressive near vertical wall bounding a glacial cirque that overlooks Owens Valley.

To the north of Mount Whitney is Mount Russell, named after a geologist who in 1889 noted that the faults along the base of the Sierra Nevada are active as based on fault scarps near Mono Lake. Such as near Mount Whitney where similar remnants of the upland surface preserved to the north can be seen along the east ridge of Mount Russell (south side of Mt. Carillon, elevation between 3900 to 4100 m) and the south flank of Mt. Hale (3900 to 4100 m).

Figure 45. Photograph of Mount Whitney, the starting point of the John Muir Trail and the highest point in the 48 contiguous states.

Cenozoic Erosion Surfaces

The cirques of the high Sierra have cut into older erosion surfaces, the highest of which was called the **summit uplands** by Lawson (1904) and described in greater detail by Knopf (1918). The summit upland plateaus were also correlated to Eocene surfaces by Matthes (1930; 1960), and are also known as the Eocene peneplain, a feature that once covered most of the Sierra Nevada block. Most of the upland surfaces are located at the summits of peaks along the divide of the range. These surfaces in part may correlate to those underlying Eocene gravel deposits described by Lindgren (1911) in the northern Sierra Nevada, but the summit areas have undergone additional erosion removing all soil profiles and were modified by frost heaving. In addition, the area near Sonora Pass and to the north have Miocene volcanic rocks deposited directly on top of granitic bedrock, leaving open the possibility of younger post-Eocene erosion surfaces. In the Pleistocene, the surfaces may have had limited glacial ice covering them, and certainly permanent snowfields. From south the north, some of the better known remnants of the summit upland include Mount Whitney (4,250 to 4,416 m) and Williamson (3,900 to 4,360 m), Mount Darwin (4,200 m), the surface at Table Mountain southeast of Lake Sabrina (3,525 m), the Four Gables (3,900 m), Mount Morgan (3,740 to 3,900 m) and plateaus at Mount Conness (3,535 to 3,760 m). Most of the summit upland plateaus are curvi-planar and gently dipping so that they do not project or line up with one another, but when viewed from a distance define accordant ridges decreasing in elevation towards the north. The lack of local alignment probably indicates gently undulating paleo-topography.

Several lower elevation surfaces, called various names by Matthes (1965), suggest periodic uplift, stability and incision, however, all are poorly dated and some may represent slope variations of a single-aged surface. These intermediate elevation surfaces, following the term used by Lawson (1904) are perhaps best called the subsummit plateaus. Subsummit plateaus are found at the Boreal Plateau (3,300 to 3,475 m elevation) to the south of Whitney, a short ways north along the JMT at the area known as Diamond Mesa (3,740 to 4,000 m), and the Dana Plateau (3,415 to 3,720 m) south of Tioga Pass. Yet further inset at lower elevations is the Chagoopa Plateau (2,740 to 3,170 m), overlooking the west flank of Kern Canyon. It was partially incised by drainages that flowed into the Kern River before being lined by 3.5 Ma lava flows of the Trout Meadows basalt (Dalrymple, 1963). Note some of the subsummit plateaus in the southern Sierra Nevada have higher elevations than the summit upland plateaus from the northern Sierra Nevada. The steep-sided valley walls of Kern Canyon were called by Lawson (1904) and Matthes (1937) the Canyon stage or surface, representing the latest interval of uplift and erosion. The lower lip of the Chagoopa Plateau stands about 670 m above the floor of Kern Canyon. The summit and subsummit plateaus taken as evidence of episodic uplift and intervening periods of erosion appears consistent with up to three slope angles of range front facets along the east escarpment (Bierman and others, 1991).

Wakabayashi and Sawyer (2001) study of stream incision of the western slope suggested that the region of summit upland plateaus surrounding Mount Whitney had possibly greater than 2,500 m of relief before the main phase (before 5 Ma) Sierra Nevada uplift. Before conclusions can be drawn, it remains to be shown that the bedrock resistance to stream erosion is uniform NS along the Sierra Nevada block, age constraints on the stream incision is sparse in the southern Sierra Nevada, and stream incision models have never been integrated with an accurate profile of range front normal fault displacements. Wakabayashi and Sawyer (2001) suggested that the Whitney paleo-relief was a relic from the Mesozoic mountain belt. Likewise, various studies suggest periods of uplift of the Sierra Nevada starting before 5 Ma, and therefore some of this older topography may be related to initial faulting of the Sierra Nevada and not remnant early Cenozoic relief in the surface of the upland summits.

Study of the range uplift by stream incision, such as those of Matthes (1930) and Huber (1981), is complex and has numerous pitfalls mainly because it is an indirect method to estimate the amount uplift from the amount of erosion. Even so, recent work on dating caverns along the western slope suggests that some oldest known caves formed before about 2.5 Ma (Stock an Anderson, 2002), which is in accord with timing of deeper topography dated by lavas in filling canyons cut in the Chagoopa Plateau. The maximum amount of local relief between canyons and **interfluves** (divides between canyons) are along forks of the Kings River. From Monarch Divide down Lost Creek to the Middle Fork of the Kings River has 1,890 m of local relief. Similarly, from Wren Peak on the Monarch Divide to the South Fork is 1,900 m of relief. To the south, maximum local relief along the Middle Fork of the Kaweah River southward to Paradise Peak is about 1,800 m. The relief from the base of the lowest Kern River at the western Sierra Nevada range front, which is at 244 m elevation to the east of Bakersfield, to the top of Mount Whitney is about 4,172 m. Note that this last estimate is the relief of the river profile from headwaters to alluvial plain, a very different geomorphic parameter than the local interfluve to canyon bottom relief estimates.

Compare the above values with that of maximum crest-scarp relief of the eastern Sierra Nevada of about 2,300 m and crest-basin floor difference of 4,200 to 4,600 m, but recall the problems of measuring the amount of uplift because of no relative datum (see Chapter 1). Some component of the eastern escarpment crest to basin floor results from subsidence of the hangingwall block. The greatest local relief along Kings River places a probable minimum for the amount of uplift. Determining the age of the interfluves and the summit upland plateaus is important for using the amount of incision to calculate uplift rates for the Sierra Nevada.

As mentioned in the introduction section on uplift, the divide of the Sierra Nevada describes a highly curved line that generally follows the major steps or jogs of the range front fault system. A second order control on the geomorphology of the divide is from headward erosion of glaciers forming large embayments or reentrants along the divide. Brocklehurst and Whipple (2002) also point out that some prominent peaks, such as Mount Williamson, lie farther eastward than the divide and have higher elevations, for which they attributed to incision cutting back the range from the line of maximum block uplift. Such a pattern of high peaks located adjacent to the divide is also well developed in the Cordillera Blanca of Peru.

Some of the lower elevation plateaus along the east escarpment, east of the divide, are located along the areas of major steps in the range front fault, such as at Coyote Ridge or Warp to the southwest of Bishop. This surface is overlain by a basalt flow dated by K-Ar method at about 9.6 Ma (Dalrymple, 1963). This age represents an upper limit on the time of surface formation, and so it remains unclear on the timing of the Coyote Ridge surface with respect to the summit upland plateaus.

The summit upland and subsummit surfaces generally lack material to be directly dated. Many have been grouped into one erosional feature, sometimes called the Eocene peneplain. Similar geomorphic surfaces in the Andes, where draping volcanic deposits allow more detailed dating of erosion surfaces cutting the Coastal Batholith, have multiple erosion events during the Cenozoic. The upland summit plateaus of the Sierra Nevada probably do not define a single erosion event or an unique individual surface. This is impart supported by the preservation of late Tertiary river channels along the western foothills, such as at Table Mountain north of Oroville, and along the San Joaquin River, where circa 9 to 9.5 Ma basalts overlie gravel beds filling channels (Lindgren, 1911; Matthes, 1960; Dalrymple, 1963). Acknowledging that the summit upland surfaces have had recent erosion removing all soil profiles and any covering deposits, it then becomes difficult to assign these surfaces to the Eocene or late Tertiary.

A View to the East

Directly below the Whitney summit is the valley containing Whitney Portal; the vertical relief from the trailhead to the summit is 1,869 m (6,133'). The JMT along the Whitney Crest is the only portion of the trail where a view into Owens Valley is obtained.

Owens Valley and the Sierra Nevada formed by normal faults along the eastern range front. The eastern side of the fault was dropped down during the formation of the valley, and the western side was uplifted, producing the mountain range. This style of faulting extends from the Sierra Nevada to the Wasatch Mountains of Utah, and region between is the Basin and Range province. At the base of the Sierra Nevada, the range front faults are more complicated than just a single structure. Looking down on the Alabama Hills, which is located just west of Lone Pine, one can see the abrupt eastern margin of the exposed granitic rock cause a linear feature. This marks the active Lone Pine fault that runs between Owens Lake and Big Pine. In addition, Owens Valley contains several buried fault strands that help accommodate the subsidence of the graben.

To the east of Owens Valley are the Inyo Mountains, mostly composed of folded Paleozoic miogeoclinal (continental shelf) limestones, Jurassic to Triassic metamorphosed volcanic rocks, and Cretaceous plutons. The Inyo-White Mountains are of comparable elevation as the Sierra Nevada, but is situated in the rain shadow of the Sierras, making it a desert range. Along the crest of the northern part of the range are groves of Bristle Cone Pines, the oldest living trees. The Inyo Mountains apparently lack multiple paleo-erosion surfaces that would be comparable to those of the Sierra Nevada, which is probably a function of the weaker bedrock and maybe a different uplift history. However, the Inyo Mountains did have at least a two stage uplift history starting in the Miocene and then rejuvenated at about 3 Ma (Conrad, 1993; Lueddecke and others, 1998).

View to the South

To the southwest of Mount Whitney one can see the distant Kern River Canyon, defining a north-south trending valley cut into the gentle Chagoopa Plateau. It marks a centerline in a wide fault zone called the proto-Kern River fault (Fig. 45); a Cretaceous, ductile, right-lateral fault zone active between 105 Ma and 85 Ma (Busby-Spera and Saleeby, 1990). This earlier phase of faulting was included in the Sierran Crest shear zone by Greene and Schweickert (1995). The term **ductile** refers to the rock having the mechanical property to flow under the right temperature and pressure conditions. To illustrate ductile behavior of rocks, imagine a bar of taffy placed half way hanging off the edge of a counter top. The taffy will stretch slowly over several days until it reaches the floor. If you took the same taffy bar and slapped it hard against the counter top it would shatter into numerous fragments, a behavior called brittle. The difference between the two cases is the applied deformation rate. Any rock under high pressure and temperature will flow like taffy in response to a slow deformation rate. This style of rock behavior is called plastic strain, and the type of rock textures produced is ductile deformation. Commonly, ductile deformation is marked by foliated rock. The proto-Kern River fault was over printed by younger faulting, and is active today as a narrow zone of right-lateral brittle deformation (Webb, 1936; Moore and duBray, 1978). This fault controlled the drainage pattern of Kern River. Valley glaciers also formed along it during both the Tioga and Tahoe stages (Matthes, 1965). The flat plateaus to either side of Kern River Canyon perhaps were glaciated by a pre-Tahoe stage. Moreover, the proto-Kern River fault was probably active throughout the formation of the Mount Whitney Intrusive Suite. What role if any this fault played in the emplacement of the plutons is unknown.

Farther south, the Sierra Nevada steadily losses elevation, and then drops steeply at the range front near the Garlock fault, which crosses the trend of the mountains at nearly a right angle. The left-lateral Garlock fault is one of major recent faults of California, moving in response to motion of the San Andreas fault. Farther to the south, the Sierra Nevada ends at the Mojave Desert (not visible from the Whitney summit).

Mount Whitney Intrusive Suite

The Mount Whitney is composed of a rock unit called the Whitney Granodiorite. The Whitney Granodiorite is the youngest pluton of the Mount Whitney Intrusive Suite (Fig. 46). This associated group of granitic plutons is of slightly younger age and is a bit larger than the Tuolumne Intrusive Suite at the north end of the JMT. From oldest to youngest, the suite is composed of the Sugarloaf Granodiorite (<88 Ma, Chen and Moore, 1982) and Lone Pine Creek Granodiorite (87 Ma, Chen and Moore, 1982; the latter pluton is exposed along the Whitney Portal road), the Paradise Granodiorite (86 Ma),

Geology of the John Muir trail

and the Whitney Granodiorite (83 Ma)- see Appendix I for sources of age data. The pattern of mineral zonation, development of foliation, and abundance of mafic inclusions all have a gradational pattern that varies away from the major pluton contacts (Fig 47). Hirt (1989) estimated the emplacement depth of the Lone Pine Creek Granodiorite at about 864 km using a method of geobarometry on the mineral hornblende. Subsequently, Hirt (2007) refined his estimate on the depth of formation to about 10 km, and furthermore related the development of mineral zonation to the gradual warming of the crust from multiple intrusions. This fundamentally returns to the perspective or field observation that large crystals develop during slow cooling of a magma. In some respects, this large-scale crystal fractionation within the nested plutons is an immense example of Bowen's reaction series.

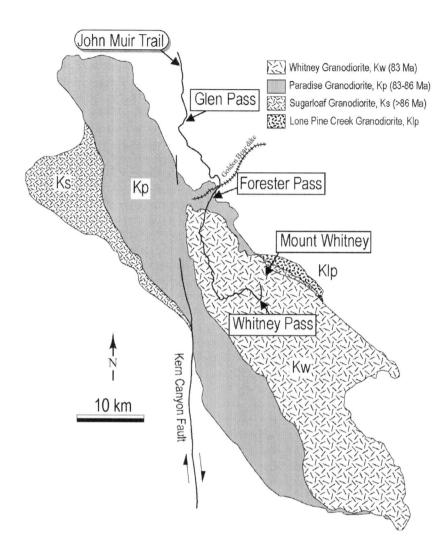

Figure 46. Geology of the Mount Whitney Intrusive Suite. Note the northwest elongate shape of the plutons and the age of the intrusions becoming younger towards the center. Geology after Moore (1981) and Hirt (1989).

Whitney Granodiorite (Kw)

The Whitney Granodiorite (Kw), 83 Ma (Chen and Moore, 1982), is a Late Cretaceous zoned porphyritic granodiorite to granite pluton. Its most distinctive feature is the abundant, rectangular, megacrysts of potassium feldspars (4-8 cm in length). The Whitney Granodiorite is the youngest pluton in the Mount Whitney Intrusive Suite (Fig. 46). Mafic inclusions are rare in this rock, and biotite is the most abundant dark mineral. Sphene is a common accessory mineral (Hirt, 1989). Overall, the rock is light colored. The pluton is subtly zoned, increasing in quartz towards the center, yet this will probably not be noticeable to the casual observer.

From the summit of Mount Whitney it will take the average backpacker two days to hike across the Mount Whitney Intrusive Suite! Imagine all this volume of rock being molten, possibly 6-8 km below the surface. Furthermore, both to the north and south of this pluton were similarly sized bodies of magma being intruded into the crust. The current and long-standing debate amongst geologist is how to provide room for all this molten rock because there had to be older rocks present before the intrusion of the pluton. The geometry of the Mount Whitney Intrusive Suite and the Tuolumne Intrusive Suite to the north, both suggest the younger plutons injected into and either cut out or shouldered aside the older plutons. In the case of the Mount Whitney Intrusive Suite, the Whitney Granodiorite intruded the pluton composed of the Paradise and Lone Pine Creek Granodiorites. Did the Paradise/Lone Pine Creek pluton pull apart to create space for the Whitney Granodiorite (tectonic emplacement) or did it get pushed aside from the intrusion of the Whitney Granodiorite (forceful emplacement)? As another option, it is possible large blocks of the Paradise/Lone Pine Creek pluton were stoped or split off from the magma chamber roof and sank through the molten Whitney Granodiorite as the pluton ascended (passive emplacement). However, blocks of the older pluton are not seen contained in the Whitney Granodiorite. In addition, there is little age difference between the plutons, so the

> Paradise/Lone Pine Creek pluton was probably still hot throughout the intrusion of the Whitney Granodiorite. When rocks are hot they generally do not break sharply, instead they tend to flow, suggesting the passive mechanism would not work. Much of the conflict between pluton emplacement models probably stems from the arguments being based on only one aspect of the intrusion process when the above-mentioned mechanisms may be occurring simultaneously.

From the summit of Mount Whitney, the trail runs south, passing Keeler and Day Needles, and then traverses by Muir peak. The near vertical chutes between the needles may be easily approach to peer down towards the east. From the summit, the trail along the jagged crest is the highest and most spectacular section of the John Muir Trail. Along this section of trail many aplite sills (horizontal sheets of intruded magma composed of quartz and potassium feldspar) cut the Whitney Granodiorite, except are poorly exposed because of talus cover. A similar sill is exposed at the trail junction (384389E, 4046702N) for the pass at Trail Crest (the sill at the junction is exposed behind the trail signs). The aplite sill is finer-grained and lighter cream colored rock as compared to the overlying more coarse-grained k-spar megacryst bearing Whitney Granodiorite. The aplite sill may be some of the very last melt to crystallize and probably contained a high percentage of superheated water when it intruded.

JMT trail log

From the Trail Crest junction, the trail descends many extended switchbacks on talus and outcrops of the Whitney Granodiorite for 4-km (2.5 miles) to Guitar Lake. The trail passes the north side of Guitar Lake, which sits in an immense amphitheater of highly jointed granodiorite. Glaciers scoured out the depression in which holds the lake. Slowly, the two stream inlets to the lake are depositing sand and gravel, and will eventually fill the lake with sediment, forming a meadow. The large wall to the south of Guitar Lake is Mount Hitchcock, and it is cut by E-W striking vertical joints in the Whitney Granodiorite. The joints preferentially weather out to form parallel blades of rock and intervening chasms. Each winter these joints are lined by ice that expands inside them, a process called **frost wedging**. The repetitive thaw and freeze of ice within fractures and joints breaks apart the rock, sending large blocks crashing down the wall, and building up the talus skirts at the base of the cliffs. Most of the debris were deposited since the last major glaciers melted. If all the talus blocks were to be counted and then divided by the number of days in 10,000 years, probably when the last glacial ice melted, the frequency of rock fall should be convincing enough not to camp on the talus fields.

The trail next descends to Timberline Lake. From the lake the trail continues down, passing several aplite dikes and sills in the Whitney Granodiorite. The outlet drainage from Timberline Lake has incised a gully 20 m (~60 feet) deep into the glacial carved valley floor since the last glacial event. Winter ice breaks apart the rock along joints and sheeting planes, dislodging the rock in the streambed to be carried away in the Spring by the higher water discharge.

The trail descends along the north side of the Whitney Creek for 2.4 km (1.5 miles), passing the side trail for Crabtree Ranger Station, and then continues westward over Tioga till (Qg- these moraines make a flat sandy section of trail). At the next trail junction (waypoint 378473E, 4046623N), the John Muir Trail merges with the Pacific Crest Trail (PCT), or vice versa. The PCT runs about 4,263 km (2,650 miles) between the borders of Mexico and Canada.

Just north of the trail junction with the PCT, the John Muir Trail ascends a prominent lateral moraine of Tahoe till (Qta). Upon descending the lateral moraine, the trail crosses a drainage, and contours around the east side of Sandy Meadow. The meadow is composed of Sherwin till (Qs) for the next 2.4 km (1.5 mile) along the JMT. The trail climbs north from Sandy Meadow to the top of a rounded ridge. To the east of the saddle, on the south side of the ridge, are weathered outcrops of the Whitney Granodiorite. From the saddle, the trail descends over the moraine crests of Tahoe till, followed by several moraine crests in younger Tioga till.

Upstream from the Wallace Creek ford, the stream incised about 15 m (45 feet) into the glacial carved valley floor. The trail climbs up from Wallace Creek over Tioga till to go around a low ridge, and then crosses Wright Creek. To the north of Wright Creek, the trail passes a bouldery Tioga stage moraine crest parallel to the elongate meadows draining to the southwest (Fig. 48).

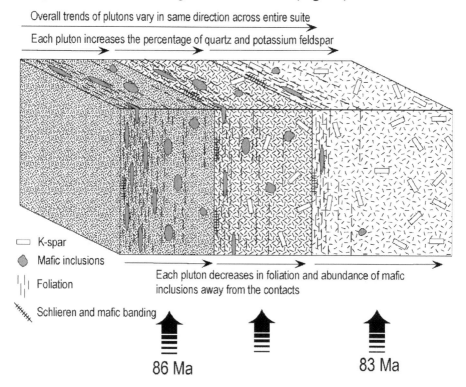

Figure 47. Diagrammatic block section showing the textural and mineralogic variations within the Mount Whitney Intrusive Suite as based on rock descriptions given in the dissertation by Hirt (1989). Sketch is not drawn to scale; the horizontal scale is in kilometers whereas illustrated textural features are in the centimeter up to a couple of meters scale as seen in the outcrop. These patterns may be explained by multiple injections of magma followed by periods of cooling and solidification. Foliation adjacent to the contacts perhaps represents flow of the magma and/or deformation from pressurization of the younger magma chamber by injection of new melt. Mafic inclusions are more elongate or stretched nearer to the intrusive contacts.

Figure 48. Boulders in the Tioga till lateral moraines located northwest of Wright Creek. View to the southeast.

Crossing Bighorn Plateau, most of which is composed of Sherwin till (Qs), the JMT passes a small lake on the east side. The Sherwin till here comprise barren exposures containing few large round boulders and the deposit lacks moraine crests. Moore (1981) mapped this unit as being older than Tahoe till, but did not speculate to which stage it belonged. I have assigned it the Sherwin stage as based on the lack of moraine crests and weathered aspect. This deposit is most likely the oldest till along the JMT and possibly in the southern Sierra.

To the north of Bighorn Plateau is Tawny Point (3,758.6 m, 12,332'), which is made of the Whitney Granodiorite. The trail gently descends to cross the contact between Sherwin and Tahoe tills at the break in slope forested by Foxtail Pines. The Foxtail Pines are a related species to the Bristlecone Pines, the oldest known living trees. The Tahoe till on the flank of Bighorn Plateau forms a lateral moraine of bouldery material (Fig. 49). The trail descends on Tioga till along the east valley wall, going over several moraine crests. Along this slope the JMT passes the unmarked trail junction for the **Shepherd Pass access (ENTRY POINT, Map 4, waypoint 376008E, 4055534N).**

Just upstream of the Tyndall Creek ford, the stream has cut down to form a low rock bank, exposing sheeted Whitney

Granodiorite. The trail gently climbs north over Tioga till (Qg) and runs parallel with lateral and medial moraines for the next 2.4 km (1.5 miles), briefly following the margin of a lateral moraine composed of Tahoe till. The Tahoe till is mostly higher on the west valley wall and it composed of numerous large boulders. The trail continues on Tioga moraine for 1 km (0.62 mile), and then over glacially carved exposures of the Whitney Granodiorite partially covered by patches of till and meadow for the next 1.2 km (3/4 of mile). This section of trail is barren, except for grass, and it is easy to imagine the Tyndall Creek glacier having just melted away leaving this newly formed landscape.

To the west of the large lake in Tyndall valley, the trail follows along the top of a narrow band of Tioga till, approaching the contact between the Whitney Granodiorite (Kw) and the Paradise Granodiorite (Kp). Right where the trail intersects the contact is a covering patch of Tioga till. A 100-meter walk to the east across flat sandy ground brings one to excellent glacial carved exposure of the contact (Fig. 50). The contact strikes east-west, and to the east its trace divides the middle of Diamond Mesa (the contact is difficult to see from a distance). Diamond Mesa defines a broad plateau of lower elevation than the summit upland surfaces found along the divide to the east.

The intrusive contact is a complex zone, about 50-m wide, and cut by aplite and pegmatite dikes and sills. Nearest to the Paradise Granodiorite / Whitney Granodiorite contact, the Whitney Granodiorite contains distinct light and dark layering called schlieren (Fig. 51).

Geology of the John Muir trail 115

Figure 49. View southward at moraine deposits at Bighorn Plateau. Bouldery material at center of photo is Tahoe stage lateral moraine (Qta), and to the left the barren smooth area is till of the Sherwin stage (Qs).

Figure 50 (lower opposite page). Photo looking east at the contact between the Whitney (Kw) and Paradise (Kp) Granodiorites. The contact is at the dead center of the photograph (arrow). I had to walk up and down the slabs to locate it. On a bright day it may all appear like the same rock. Sometimes it helps to use sunglasses, or not to use them, depending on the lighting, to see the difference in color between the units.

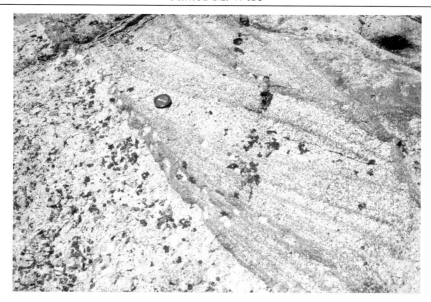

Figure 51. Detail of the contact between the Whitney and Paradise Granodiorites. The Whitney Granodiorite is on the left side and is more coarse-grained. Large white crystals are megacrysts of potassium feldspar, which are oriented with the long sides of the crystals parallel to the surrounding layers. Bands of black minerals define the schlieren structure, which here dips steeply and has curved bands. Note truncation of layers in the upper right corner of the photo indicates that rock to the left formed later. Lens cap for scale.

On the Origin of Schlieren

The compositional layering called **schlieren** has a pattern similar to cross-bedding commonly formed in sedimentary rocks by grain settling out from a current. The process causing schlieren as of yet is not completely understood, however, some geologist infer magma flow in a pluton as the mechanism to produce the layering. Schlieren structure has been described from plutons around the world, and occasionally is developed within dikes (Reid and others, 1993). Disrupted and deformed inclusions have also been called schlieren, but the more restricted usage of Bateman (1992) excludes these types of inclusions. Most schlieren define alternating bands of mafic and felsic minerals, commonly with graded or sorting of the grain sizes, in parallel or converging layers.

Because schlieren are partially composed of concentrated mafic heavy minerals, commonly hornblende, biotite, zircon, sphene, and apatite, perhaps these minerals separated from the rest of the pluton by gravitational settling. However, this may be difficult to do because the viscosity, the ability to flow, is relatively low for quartz-rich melts when containing abundant crystals. The density of the mafic minerals, such as hornblende, as compared to the density and viscosity of the surrounding magma may not be enough to mechanically sink a crystal. Also, unlike cross-bedding in sediments, the orientation of layering in schlieren is highly variable with respect to vertical, suggesting that in most cases gravity was not an important control. Also important to keep in mind when examining the banding in schlieren is the character of the surrounding rock; if the rock is foliated and the mafic inclusions are flattened or stretched it is likely that the angles of cross-bedding and truncation in the schlieren have decreased by deformation.

Commonly layers in schlieren have been truncated or eroded, a feature that Bateman (1992) called unconformities, borrowing the term from a type of sedimentary contact. Other examples have channel forms cut into the earlier formed rock. These structures imply alternating periods of mineral deposition/precipitation and erosion/dissolution and probable magma flow. To form one class of schlieren, it therefore appears that the mafic crystals must be mechanically segregated out of the melt, either transported by settling or relocated through shearing and pressure gradients. The second process of shearing has been used to explain phenocryst distributions in dikes, and mechanical sorting of particles in a liquid undergoing shear has been reproduced in scaled models. The flow velocity of a fluid increases away from the contact with a solid interface. Particles entrained respond to the pressure gradient of this differential flow, or shear, by the larger particles, which have more surface area and therefore frictional interaction with the surrounding fluid, moving towards the regions of high flow rate. The margins of plutons are more likely to record flow of the magma, and therefore shear along the fixed wall rock, as the pluton is emplaced, and perhaps explains why schlieren are more common near major intrusive contacts.

Above I have listed two processes, mineral settling and mechanical sorting, which have been used to explain schlieren. Perhaps finer subdivisions are needed to describe the wide range of structures that are included in schlieren. Schlieren may also be related to magmatic banding and cockcomb textures that represent accumulation of minerals at a crystallization front or the solid-liquid interface accompanied with magma flow.

Paradise Granodiorite (Kp)

The Paradise Granodiorite, Map 5, 84-86 Ma (Chen and Moore, 1982), is a porphyritic rock containing 1-3 cm long potassium feldspar crystals that have inclusions of biotite and hornblende along the internal growth zones. The average mode is 22.5% quartz, 22.0% potassium feldspar, 43.6% plagioclase, 6.7% biotite, 3.5% hornblende, and 1.8% accessories (Moore, 1963), giving an overall composition of granodiorite. The rock is darker colored than either the Bullfrog Pluton (which lies to the north) or the Whitney Granodiorite (Fig. 46). Mafic inclusions are common, and Hirt (1989) reported that locally hornblende crystals are up to 2-cm long.

North of the intrusive contact near the JMT, numerous aplite dikes originating from the Whitney Granodiorite cut the Paradise Granodiorite. The dikes decrease in abundance steadily northward away from the contact. The trail passes on the western side of two large lakes, climbing over till, fords Tyndall Creek just south of the outlet to a small lake, and then ascends to Forester Pass via impressive manmade cuts in the steep cliffs built in 1932. On the ascent to the pass, the Paradise Granodiorite contains a fair amount of mafic inclusions exposed along the trail. These darker blobs were once a molten rock of different composition than the surrounding rock and were mixed before crystallization of the pluton.

Nearing the upper section of the climb, the prominent chute the trail traverses across marks the position of a fault, which cuts through Forester Pass (the chute is frequently snow filled). The high-angle brittle fault has completely fractured and shattered the rock (Fig. 52), producing a zone of weakness acted upon by erosion, forming the ideal location for the pass. The rock at the pass is ground up by the fault into a breccia (an angular clast in matrix shattered rock). The fault rocks were altered to pink and greenish colors, imparted by the hydrothermal fluids that had moved through the fault zone and chemically reacted with the Paradise Granodiorite. The green minerals may be chlorite and epidote.

Forester Pass (4,017 m, 13,180', Map 5, waypoint 377389E, 4061647N)

Forester Pass is highest pass along the JMT (excluding Trail Summit), and it marks the boundary between Kings Canyon National Park to the north and Sequoia National Park to the south. To the east of the pass, along the ridge, is Junction Peak (4,232.8 m, 13,888'). On the opposite side of the cirque to the west and south of the pass is Caltech Peak (4,215.8 m, 13,832'). Both peaks are composed of the Paradise Granodiorite.

The view to the north looks down the U-shaped Bubbs Creek Valley, and down onto rock glacier deposits (Qr) lining the floor at the head of the cirque. To the south, Tyndall Creek glacial valley is lined by moraine deposits of the Tioga and Tahoe stages. The valley joins the Kern Canyon drainage farther to the south. The east valley wall of Tyndall Creek forms Diamond Mesa. To the south of Diamond Mesa is the distance the low east valley wall is topped by barren Bighorn Plateau (Fig. 53). Far to the southwest are the dark Kaweah Peaks, composed of metamorphosed volcanic and marine sedimentary rocks.

Figure 52. Photograph of Forester Pass viewed from the south. The notch in the pass is from weathering of a high-angle fault zone that contains crushed rock that is called breccia.

Figure 53. Photo looking south from Forester Pass. The contact between the Whitney Granodiorite (Kw) and the Paradise Granodiorite (Kp) exposed between the two partially frozen lakes.

Geology of the John Muir trail 121

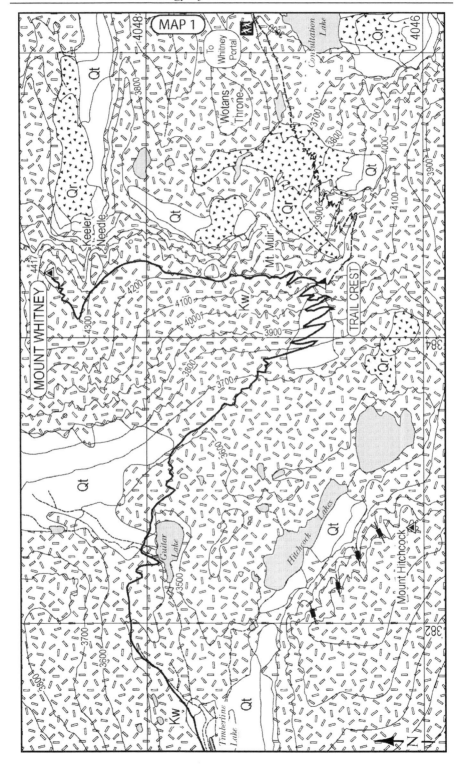

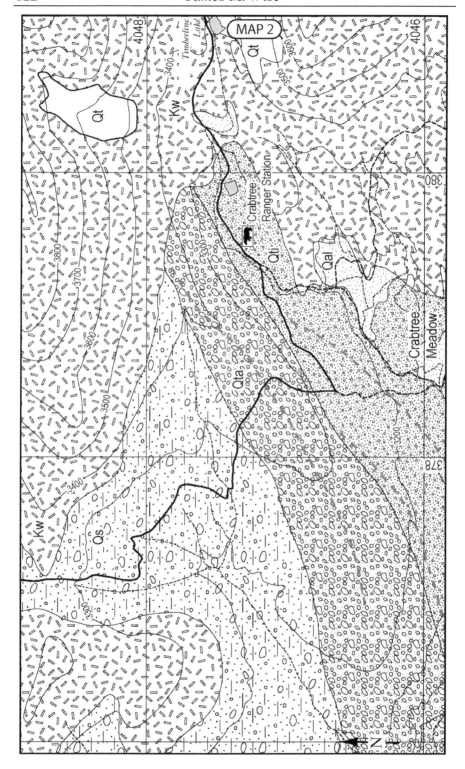

Geology of the John Muir trail

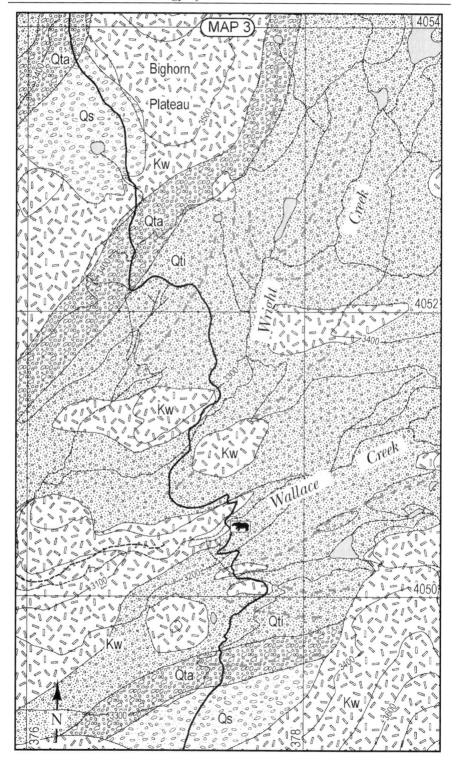

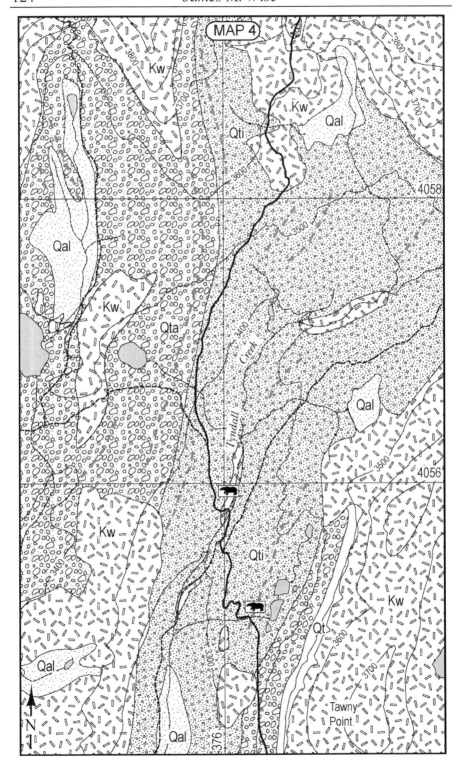

Geology of the John Muir trail 125

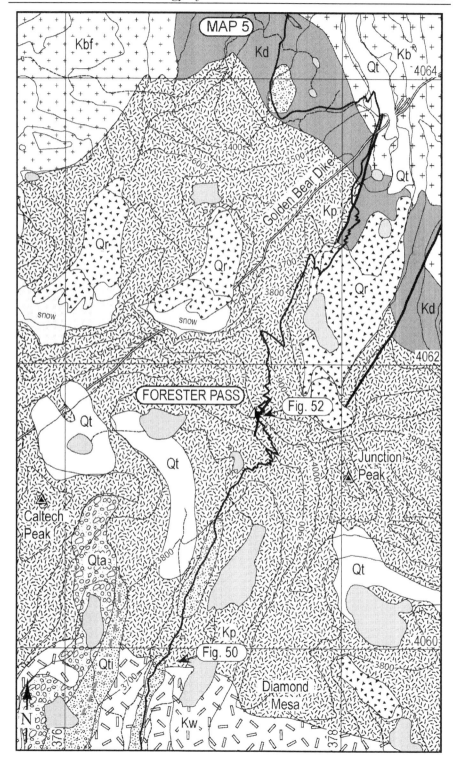

CHAPTER 3

FORESTER PASS TO GLEN PASS

Access: Kearsarge Pass, and Bubbs Creek from Cedar Grove.
Distance: 16 km (10 miles).
Maps: 5-7.

This segment of the trail passes through spectacular rock glacier deposits, crosses over mafic plutonic rocks, the Golden Bear Dike, the Bullfrog Pluton, and a mafic intrusive complex at Glen Pass (Fig. 54). The Mount Whitney Intrusive Suite is left behind for slightly older rocks.

From the low notch in the cirque that makes Forester Pass (4,017 m, 13,180') the trail diagonals down to the north (commonly over snow patches) on an elongate glacial-carved ridge line. In blasted sections of the trail, the Paradise Granodiorite is well exposed and includes k-spar megacrysts up to 3 cm in length. Once the trail descends into the glacial valley it switchbacks down and across rock glacier deposits (Qr) (note the jumbled, blocky appearance). The JMT crosses onto diorite plutonic rocks (JTrm); the rock is dark gray, medium-grained, and is cut by numerous aplite sills originating from the Bullfrog Pluton. On the ridge to the east of the trail, the darker colored diorite is in clear intrusive contact with the Bullfrog Pluton. The Bullfrog Pluton injected into the diorite. The trail traverses on this JTrm unit for 1.2 km (3/4 of a mile). Although Moore (1981) inferred a Jurassic age for the diorite, this unit may instead be Cretaceous and close in age to the Bullfrog Pluton as based on the similar rocks near Glen Pass and Onion Valley.

Upon reaching the ford of a stream, to the east is a north-striking fault that offsets the mafic plutonic rock and the Golden Bear Dike. The dike is exposed right along the trail and in the ridge to the east of the trail. Locating the dike takes a few minutes of examination, because the dike and the rock to either side is of similar color and texture. Look for a general texture change accompanied by a slight change in the rock color, and recall that a dike represents a vertically oriented intruded panel of magma.

Golden Bear Dike

The Golden Bear Dike, 80 Ma (Chen and Moore, 1982), is the youngest granitic intrusion along the JMT (Fig. 54, Map 5). The average width of the dike is ten meters, and its length is about 14 km. Relatively speaking, this is a very long dike. It contains coarse-grained quartz, feldspar megacrysts, and biotite in a fine-grained matrix of quartz and feldspar. The dike contacts with the county rock are sharp, and the margins of the dike are finer-grained, a texture suggestive of quenching the magma lining the dike against colder wall rock. Along the contact is minor mineralization of galena (PbS), chalcopyrite ($CuFeS_2$), and chrysocolla ($(Cu, Al)_2H_2Si_2O_5(OH)^4*nH_2O$). Most metallic minerals are products of late stage fluids that come from the crystallization of magmas. The minerals along the borders of the Golden Bear Dike are of insufficient quantity to be economically mined. The dike has a slight greenish hue that subtly contrasts the slight pinkish color of the Bullfrog Pluton. The dike is exposed on the steep ridge to the east of the trail where it forks into two upward-pointing, finger-like structures. Hirt (1989) inferred the formation of the dike as being related to the Whitney Granodiorite. The ENE-striking dike indicates a paleo-stress direction that is generally compatible with that which formed the Kern Canyon fault to the south.

From the Golden Bear Dike, the trail executes a short descent and nicks the margin of the Bullfrog Pluton. It then re-enters the diorite, turns to the west, and circles around the south end of a tarn on flat meadow, which has protruding glacially carved slabs of diorite (JTrm) partially covered by numerous scattered glacial erratics. The view northwestward down Bubbs Creek Valley defines the classic U-shaped valley profile marked by a distinct trim line high on the north valley wall (Fig. 55). The trim line marks the upper limit to the Tioga stage glacial ice carving at the canyon walls. The trail descends and intersects the northern contact of diorite (JTrm) with the Bullfrog Pluton (Kb) along the east valley wall of the Bubbs Creek (Fig. 56, Map 6). Looking to the west across the creek to the opposite valley wall, the diorite clearly stands out in color that contrasts with the surrounding granodiorite. The spatial and temporal relationship between granodiorite and diorite is a common phenomenon in the Sierra Nevada batholith, as seen from the larger bodies crossed here and at the smaller scale of diorite inclusions within the granodiorite plutons. The separation and later remixing of the granodiorite and diorite is part of the magma evolution from the original parent melt. Magma evolution involves several steps or processes. The large zoned plutons record both crystal fractionation and renewed melt intrusion. The intermingled diorite in granodiorite is a relict of an earlier process of the melt history. Next, the trail descends along the southeast valley wall, and passes on the western side of Center Peak (3,889 m, 12,760'), which defines a large arete.

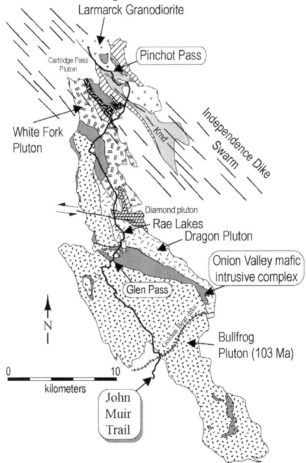

Figure 54. Geologic map of the Bullfrog Pluton and nearby rock units. The geology was modified after Moore (1963). Age data on the Bullfrog Pluton is from Chen and Moore (1982).

Geology of the John Muir trail

Figure 55. View northward down Bubbs Creek. Bouldery material in the foreground is Tioga stage till. Glacial trim line (TL) is above the tree line and at the base of the barren ridge.

Figure 56. Contact between a diorite unit (dark gray) and the Bullfrog pluton (light gray) exposed along the John Muir Trail. View is to the east (Map 6). A backpacker stands just to the left of the contact, blending in with the white boulders.

> **Bullfrog Pluton (Kb)**
>
> The Bullfrog Pluton, Maps 5-9, 103 Ma (Chen and Moore, 1982), is composed of quartz monzonite to granite, is coarse-grained, and has a slight pinkish hue imparted to it from the potassium feldspar. This rock unit was named after Bullfrog Lake by Moore (1963). The main pluton mass is 37 square km in map area (24-km long and 9.7-km wide). It is elongate and trending in a northwest direction (Fig. 54). The average mode is 25.0% quartz, 37.3% perthite (a type of k-spar), 34.7% plagioclase, 1.8% biotite, 0.2% hornblende, and 1.0% accessories (Moore, 1963). It contains almost no mafic inclusions, and is difficult to distinguish internal structures or foliation. Along most of its margins the pluton sent out numerous sills. Some of its contacts were cut by post-intrusion faults of minor offset. The Bullfrog Pluton is associated with the nearby Dragon Pluton (103 Ma) and the Independence Pluton (112 Ma; Chen, 1977), and all three may together comprise an intrusive suite. Fin Dome, King Spur, Mount Clarence King, Kearsarge Pinnacles, Charlotte Dome, and Independence Peak are composed of the Bullfrog Pluton.

After crossing the drainage below Golden Bear Lake, the trail passes opposite of a rock glacier perched in a tributary valley on the south valley wall (Fig. 57). Across from the rock glacier, the trail descends to a small, low profile, post-Tioga stage sandy alluvial fan that is outlined on map 6. Valleys of the Sierra Nevada are commonly lined by various materials, such as till, fluvial gravel and sand bars, talus, alluvial fans, sandy deltas building at the lake inlets, and fine muddy sediment collecting at the lake bottoms.

The trail gradually descends along the east side of Bubbs Creek. Along the east valley wall are the Kearsarge Pinnacles, composed of the Bullfrog Pluton. The base of the Kearsarge Pinnacles is skirted with talus. To the southwest is East Vidette Peak; on its lower-angled flanks the granitic rocks are highly sheeted (a fracture pattern related to pressure release). The trail circles around Vidette Meadow, and then fords a stream that drains from Bullfrog Lake. About 150' east of the trail junction for **Cedar Grove (ENTRY POINT, Map 7)** and just south of the trail are slabs of the Bullfrog

Pluton that have a magmatic foliation oriented 354°, 80° NE, as defined by biotite and aligned mafic inclusions. The trail ascends from Bubbs Creek junction at Lower Vidette Meadow along the tributary from Bullfrog Lake, crossing over till and talus. Tioga till lines the valley floor of the Bullfrog Lake drainage.

Figure 57. View south at a rock glacier deposit in tributary to Bubbs Creek. Rubble in center of the photo comprises the rock glacier.

Glacial Escape or Piracy?

The Kearsarge ridge formed by parallel glaciers running down Bubbs Creek and Charlotte Creek. The gap through which Bullfrog Creek flows was identified by Knopf (1918) as an area of glacier piracy, an idea borrowed from the process of headward stream erosion called stream piracy where one drainage captures the flow of an adjacent drainage. The glacier that flowed across the Bullfrog Lakes area probably continued westward down to Charlotte Lake. As the glaciers bounding Kearsarge ridge lowered the divide, the Bullfrog glacier overflowed into the adjacent Bubbs Creek glacier, creating an ice spill way. Downward cutting of the spillway, while allowing the Bubbs Creek glacier to capture the flow from the Bullfrog glacier, is a distinct process from true stream or glacier piracy where the flow being captured is robbed by headward erosion of the pirate stream.

The JMT makes a short, steep, northward climb out of the Bullfrog Lake drainage to reach the junction for Charlotte Lake. Taking the north fork, the trail ascends a little to the junction for **Kearsarge Pass (ENTRY POINT, Map 7, waypoint 374058E, 4070474N)**. The trail climbs the slope east above Charlotte Lake, parallels an elongate east-west trending patch of talus (Qt) along its north side, and then passes two small lakes on the western side. The lower lake is blocked by a dam of rock glacier debris. The JMT crosses the sharp contact between the Bullfrog Pluton (Kb) and mafic intrusive rocks (Kom), and then ascends to Glen Pass. The contact arcs across the upper ¾ part of the cirque, a dramatic exposure between two different compositions of rock, yet both may have been molten at the same time.

Figure 58. View northward from Glen Pass looking at 60 Lakes Basin. Running diagonal from lower right (starting just north of the foreground lakes) to upper center is a fault scarp with the north side apparently up (dashed line). This structure offsets intrusive contacts, demonstrating it to be post Late Cretaceous, however, its timing in comparison with the Cenozoic deformational history of the Sierra Nevada is unknown. The fault is brittle and restricted to a narrow zone.

Glen Pass (3,650.7 m, 11,978', Map 7, waypoint 374064E, 4072269N)

The mafic intrusive rocks at Glen Pass form a body of hornblende diorite related to the **Onion Valley layered mafic complex** (Sisson, 1991; Sisson and others, 1996). The mafic sills at Onion Valley have chilled margins (a very fine-grained cooling texture), are up to 1.5-m thick, and are separated by thin layers of lighter colored granodiorite (Sisson and others, 1996). Associated with the sills are concentrations of mafic minerals called cumulates. The cumulate minerals are composed of olivine, pyroxene, and hornblende. These minerals are relatively heavy, allowing them to sink through the once molten rock and settle on the magma chamber floor. The mafic rocks were dated at 102 Ma (Saleeby and others, 1990) and 92.1 Ma (Coleman and others, 1995). Cross cutting relationships reported by Sisson and others (1996) indicate the mafic complex was forming at the same time as the Bullfrog Pluton (103 Ma), suggesting the 92.1 Ma age determination may be incorrect or additional younger diorite intruded into the mafic complex. Mafic intrusive complexes, such as at Glen Pass, are relatively rare in the current level of exposure in the Sierra Nevada batholith. These intermingled mafic and felsic bodies perhaps represent the lower part of plutons and were important in the compositional evolution of the magmas (Frost and Mahood, 1987; Coleman and others, 1995; Sisson and others, 1996; Coleman and Glazner, 1997). Similar mafic complexes may underlie most of the larger plutons of the Sierra Nevada.

From Glen Pass, the view northward overlooks Rae Lakes basin, Fin Dome, and the valley beyond that drains into the Middle Fork of the Kings River. Farther to the north, rising above the valley, the reddish rocks are at Mount Pinchot, a package of metamorphosed marine sedimentary rocks that the JMT passes over. Mount Clarence King stands prominently to the northwest. At the northern base of Glen Pass, rock glacier deposits (Qr) surround the upper lakes of the basin. A prominent EW-striking fault is clearly visible from the pass (Fig. 58). The JMT crosses this structure, but up close it is difficult to recognize. To the south, the previously described Kb/Kom contact is conspicuously exposed just below the pass, cutting aretes both to the east and west.

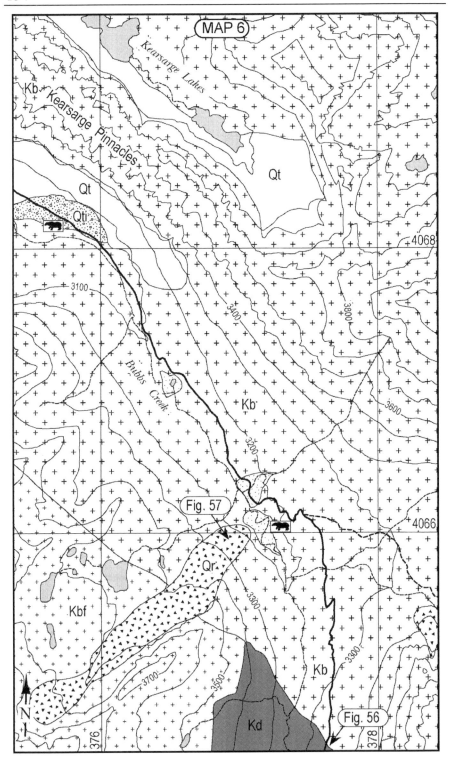

Geology of the John Muir trail

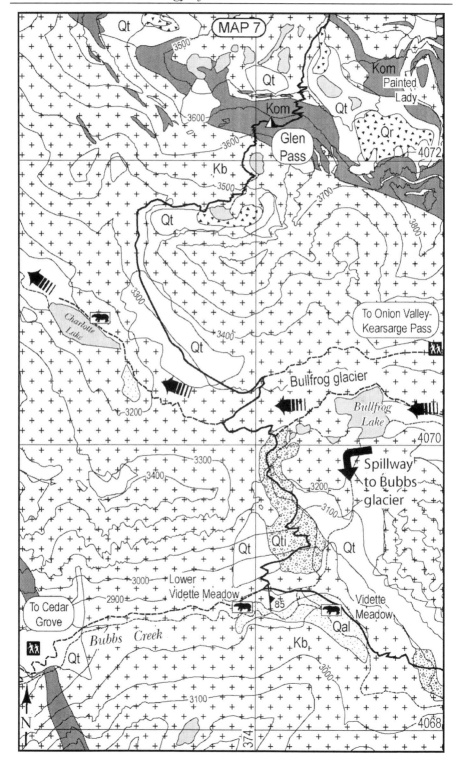

CHAPTER 4

GLEN PASS TO PINCHOT PASS

Access: Kearsarge Pass, Bubbs Creek from Cedar Grove, and Woods Creek
from Cedar Grove.
Distance: 25.3 km (15.7 miles).
Maps: 7-10.

This section of the trail traverses the Onion Valley mafic complex, the northern lobe of the Bullfrog Pluton, several older and deformed plutons, screens of metamorphosed sedimentary rocks, the Jurassic Independence dike swarm, and the Cretaceous Lamarck Granodiorite. From Glen Pass to Pinchot Pass the trail is in the watershed belonging to the Kings River.

Glen Pass (3,650.7 m, 11,978') is composed of a mafic intrusive complex (Kom), which continues to the east to comprise the Painted Lady Peak (3,695.8 m, 12,126'), and extends eastward to Onion Valley. On the northward descent from Glen Pass the trail crosses numerous thick sills of aplite intruded from the Bullfrog Pluton (Kb) into the mafic intrusive complex (Kom). However, most of the switchbacks on the north side of Glen Pass are on talus. The architecture of the rock passed over by the JMT is similar to that in the cliffs on the north face of the Painted Lady. The trail bypasses several moraines that block in lakes, crosses into Kb and then back into Kom, and descends along the south side of a drainage. The trail switchbacks down an EW-striking fault (minor offset and narrow zone of brittle deformation) that juxtaposes Kom against Kd, and then travels over the Dragon Pluton (Kd) to Rae Lakes (Map 8).

Over looking Rae Lakes from the south is the prominent peak called the Painted Lady, which is mainly composed of Kom, but is also intruded by lighter colored, thick sills of Kb (Fig. 59). The trail continues north along the west side of Rae Lakes to the trail junction for Sixty Lake Basin. Taking the east fork, the trail passes between the two southern Rae Lakes along an isthmus (here much of the rock is crumbly or deteriorated). The trail circles around the east side of the lakes, then starts the descent down the valley. To the west, Fin Dome rises prominently above the valley. It is composed of the Bullfrog Pluton and near its eastern base is the north-south striking contact with the Dragon Pluton (Kd). Fin dome is more of an arete than a dome, and its form was probably controlled by joints in the rock.

To the west of Fin Dome lies Mount Clarence King, which is named after the first director of the U.S. Geological Survey. King had organized the massive work of a geologic survey of the Fortieth parallel of the United States of America. This study, which took seven years of research, was published in seven volumes during 1870's. King's early career began as an assistant to Josiah Whitney on the Whitney Survey. He wrote up several of his climbing experiences from the Whitney Survey in a book called *Mountaineering in the Sierra Nevada* first published in 1872.

The trail intersects a minor fault contact, striking east-west, that offsets two units within the Dragon Pluton (Kd and Kdi) with apparent left-lateral slip. Unfortunately, the fault is not well exposed near the trail. The trail continues through rocks of the Dragon Pluton (Kd), Diamond Pluton (Kdi), and then back over the Dragon Pluton (Kd) while going around the east side of Arrowhead Lake. Across these contacts, the ground is covered by a patchwork of forest, meadows, and till. It crosses the contact with the White Fork Pluton (Jwf), and then fords the outlet to the lake going over Kd.

The trail traverses around the west side of Dollar Lake, over a shelf of blasted rock composed of a sliver of Paleozoic (?) calcareous hornfels (Pzch), a metamorphosed sedimentary rock, contained in the Dragon Pluton (Kd).

Diamond Pluton (Kd/Kdi)

The Diamond Pluton, Maps 8-9, has a map surface area of two square kilometers (1.3 square miles) and an irregular shape. It is older than the Dragon Pluton (103 Ma) and younger than the White Fork Pluton (>156 Ma). It has a similar composition as the Bullfrog Pluton. The average mode is 26.2% quartz, 43.3% k-spar, 25.0% plagioclase, 2.4% biotite, 0.9% hornblende, and 2.2% accessories (Moore, 1963).

Figure 59. View south of Painted Lady peak and the interlayered sills of diorite (dark gray) and the Bullfrog Pluton (light gray). The southern Rae Lake is in the foreground.

White Fork Pluton (Jwf)

The White Fork Pluton, 156 Ma (Chen and Moore, 1982), is named for the White Fork of Woods Creek. The pluton is 12.7 square km in map area, and 12.9 km (8 miles) long (Moore, 1963). It varies in composition and texture (the average composition is granite); much of the rock is sheared, some to a metamorphic rock type of gneiss or schist! Anastomosing, or braided-shaped, thin shear zones are common, as are mafic lenses, and pods of epidote (Fig. 60). The northern part of the pluton is strongly foliated. The average mode is 26.4% quartz, 19.5% k-spar, 42.1% plagioclase, 9.2% biotite, 1.9% hornblende, and 0.9% accessories (Moore, 1963). The quartz and biotite grains in the rock form elongate aggregates. It is older than the Independence Dike Swarm (IDS), and was intruded by the Granodiorite of Cartridge Pass, the Bullfrog Pluton, Baxter Pluton, and the Dragon Pluton.

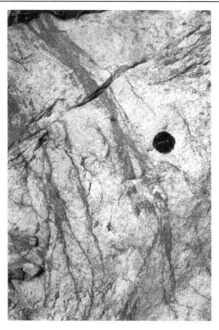

Figure 60. Epidote veins in sheared White Fork Pluton. Note abundant the fractures! This is not a pristine granitic rock. Lens cap for scale.

The trail descends northwards along the west side of the South Fork and Dollar Lake. Near the outlet of Dollar Lake, the Kd/Jwf contact makes a complex zone where Kd has ripped out chunks of Jwf and drug them along the contact zone (Map 8, waypoint 374563E, 4077261N). Abundant angular igneous inclusions, ranging from 1" to 20 feet in diameter, are composed of several different textured and colored granodiorites, and rarer meta-sedimentary rock exposed in a glacial polished slab right next to the JMT (Fig. 61). Do these blocks contained in the Kd granitic rock represent sinking of stoped material or clasts dragged upward during pluton ascent? Just to the north of this location is the trail junction to the **Baxter Pass (ENTRY POINT, Map 8, waypoint 374586, 4077296N).**

The trail descends northward and gently curves to the west. A prominent hill to the southwest has thick, nearly black, mafic dikes cutting the White Fork Pluton. The dikes are part of the Independence Dike Swarm.

Where the trail leaves Tioga till (Qg) it crosses rock of the White Fork Pluton (Jwf) and then encounters a body of mafic plutonic rock (Jm). The trail shortly descends over small intrusive body of granodiorite, unit Jwf. The trail then crosses back into Jm; the contact is just after a tributary (waypoint 373065E, 4080541N). The trail crosses two streams, then descends over talus (Qt) and the Jm/Kb contact.

The trail descends over Tioga till (Qg) along much of the valley floor. Near the convergence of the valley with the Woods Creek drainage, the trail crosses river gravels with marked linear levees made of rounded boulders. The JMT then crosses Woods Creek by a suspension bridge (any guesses on how long this structure will last with the northern footing in the riverbed?). At the north side is the trail junction for **Cedar Grove (ENTRY POINT, Map 9, waypoint 371802E, 408161N)**.

From the Woods Creek crossing, the trail ascends along the north valley wall, first over talus, then over outcrops of the Bullfrog Pluton (Kb). This portion of the Bullfrog Pluton is a northern elongate extension of the main mass (Fig. 54, map 9). This northern mass is 11.3 km (7 miles) long and 2 km (1 1/4 mile) wide. Next, the trail runs over a zone of inter-tongued Jm and intrusions of the Bullfrog Pluton. Jm is medium-grained, dark gray Jurassic diorite; best exposed in cut banks of Woods Creek and near the west contact of the unit. The trail goes over a lens-shaped intrusion of the Bullfrog Pluton, then returns to mafic plutonic rock (Jm). The trail climbs to the ford at White Fork Creek where it crosses the contact (waypoint 372403E, 408241N) with an unnamed mafic pluton (Jm) and the White Fork Pluton (Jwf).

The trail climbs and crosses a northwest-trending screen of metamorphosed sedimentary rocks. A screen is a body of rock sandwiched between plutonic rock. The screen's west side is made of pelitic hornfels (Pzph), the middle portion is composed of biotite schist (Pzbs), and the east side is composed of calcareous hornfels (Pzch). Hornfels is a name for contact metamorphosed, fine-grained, fabricless rock. Pelitic hornfels contain a high amount of aluminum whereas the calcareous variety is calcium rich. All these rocks were once deposited in a marine environment. Fossils have not been found in these rocks, so their age is unknown, except that they are older than the Jurassic plutons. The trail leaves the screen going into the Twin Lakes Pluton (Jl). After the trail crosses a septa of Pzch, it crosses over part of Twin Lakes Pluton (Jtl), and then over the Tinemaha Granodiorite (Jtn). Nearby, the trail comes to the junction for the **Sawmill Pass trail (ENTRY POINT, Map 10, waypoint 375323E, 4084768N)**. It again crosses the contact between Jtn and Jtl before reaching Twin Lakes. West of Twin Lakes are spectacular exposures of the mafic Independence Dike Swarm hosted in Jtl (Fig. 62). These two rocks of different compositions were intruded into the crust and cooled underground at different times.

To the west of Twin Lakes, the JMT crosses an interesting outcrop of metamorphosed conglomerate (waypoint 375509E, 4085470N). The meta-conglomerate contains rounded cobbles of quartzite in a quartz-rich sandy matrix (Fig. 63). While most of the stratagraphic section in the screens represent fine-grained, perhaps deep water, marine deposits, the conglomerate beds require a source area that had some topographic relief to shed the detritus. The cobbles were deformed and flattened- the timing of the deformation pre-dates the Jurassic since the IDS cuts the meta-conglomerate. A bed of marble adjacent to the meta-conglomerate, south of the trail, has epidote and garnet crystals, products of contact metamorphism.

Figure 61. Angular clasts of various composition rocks contained in the Diamond Pluton near its contact with the older Jurassic White Fork Pluton. Outcrop is north of Dollar Lake and west of the JMT. Center right of photo is my field book for scale, which is 19 cm long.

Figure 62. Glacier slab exposure of the mafic dikes cutting granitic rock. The dikes are part of the Independence Dike Swarm and are very numerous to the southwest of Twin Lakes. The dikes are striking northwest and have a vertical dip.

Figure 63. Outcrop of metamorphosed conglomerate bed in a screen to the northwest of Twin Lakes, which is crossed by the John Muir Trail. The best exposures are a few meters to the east of the trail. Age of this formation is unknown. Mechanical pencil for scale.

Tinemaha Granodiorite (Jtn)

The Tinemaha Granodiorite (Woods Lake mass), 165 Ma (Chen and Moore, 1982), is 18.5 km (11.5 miles) long, and 19 square km (12.4 square miles) in area. The average mode is 23.6% quartz, 21.8% k-spar, 41.8% plagioclase, 8.2% biotite, 2.8% hornblende, and 1.9% accessories (Moore, 1963). This rock body has variable compositions of monzodiorite, quartz monzonite, and granite. The Independence Dike Swarm extensively intruded it. Moore (1963) correlated this exposure to the Tinemaha Granodiorite of Bateman (1965) in the Big Pine quadrangle map.

Independence Dike Swarm (IDS)

The Independence Dike Swarm marks a zone of northwest-striking dikes (Fig. 54) extending from the Mojave Desert through the Inyo Mountains, and into the eastern Sierra Nevada (Moore and Hopson, 1961). The dikes range in composition from diorite to granite porphyry. The dike swarm was originally dated at 148 Ma (Chen and Moore, 1979) by uranium-lead analysis of the mineral zircon. However, dikes previously thought to belong to IDS were dated as Late Cretaceous (Coleman and others, 2000), adding uncertainty to which ones are Jurassic and which ones are younger. The dikes intruded 50 million years apart and yet lie in the same orientation and located in the same area, suggesting a deep weakness in the crust. The dikes indicate regional extension of the crust occurred in the Jurassic, producing parallel fractures perpendicular to the direction of crustal tension. The IDS also provides an useful tie point for looking at post-Jurassic regional faults in the Sierra Nevada, such as the Mojave-Snow Lake fault of Lahren and Schweickert (1990) that was completely engulfed by the Sierra Nevada batholith.

Near Twin Lakes, most of the dikes were cut by joints that are filled by epidote. The wall rock adjacent to these fractures is bleached white, indicating hydrothermal fluids moved through the rock. Many of the dikes are composites of more than one composition (Carl and others, 1997), implying the source melt for the dikes was differentiated (segregated) into several compositions during their upward migration through the crust.

The dikes dilated by oblique opening, having a component of left-lateral motion, immediately obvious from piecing together the dike margins. Glazner and others (1999) discussed the left-lateral shearing as being Jurassic in age, but they did not describe the relative motion for the younger Cretaceous dikes. Mafic dikes cutting large calc-alkaline suite magmatic arc is not an unique feature to the Sierra Nevada. The Coastal Batholith of Peru contains similar swarms of mafic dikes that parallel the axis of the batholith. The dikes represent tension in the crust, the growth of fractures and faults perpendicular to the direction of tension, and upward migration of melts along these crustal-scale weak zones. Ultimately, large dike swarms can be related to the state of motion of the major tectonic plates, and possible to the dip angle of the subducting ocean slab beneath the continental margin.

Twin Lakes Pluton (Jtl)

The Twin Lakes Pluton, Map 10, formed approximately between 165 Ma and 81 Ma, and is only composed of granite mixed with darker inclusions of diorite. This is a small pluton, about two square kilometers in map view. It is medium-grained and slightly altered, and it is crosscut by numerous micro-faults. The Twin Lakes Pluton is younger than the Tinemaha Granodiorite, and older than Cartridge Pass Granodiorite. The Twin Lakes Pluton is mixed with Tinemaha Granodiorite and diorite along the JMT in the southeast part of the intrusion. The average mode is 22.7% quartz, 34.1% perthite, 40.6% plagioclase, 2.1% biotite, and 0.5% accessories (Moore, 1963). The unit is intruded by the IDS. Part of the pluton, unit Jtld, was mapped by Moore (1963) as containing or mixed with darker granitic rocks. The igneous rock to the right of the conspicuous meta-sedimentary screen on the north face of Mount Cedric Wright towering over the southern skyline is part of the Twin Lakes Pluton.

McDoogle Pluton (Kmd)

The McDoogle Pluton, Maps 10-11, ~95 Ma (Mahan and Bartley, 2000), is a dark, coarse-grained, granodiorite marked by a strong NW-striking fabric. The pluton is 14.5 km (9 miles) long, extending from Mt. Wynne to Pinchot Pass. It is named after Mt. McDoogle (3,771.1 m, 12,373'), which is north of Woods Lake. The average mode is 12.4% quartz, 17.5% k-spar, 46.6% plagioclase, 12.1% biotite, 9.1% hornblende, 0.5% pyroxene, and 1.8% accessories (Moore, 1963). The pluton is unusual because up to 60% of the rock is composed of tabular dioritic inclusions near Pinchot Pass, securing it as the most inclusion rich pluton crossed by the JMT (Fig. 64). The fabric defined by the flattened inclusions has been locally folded at a centimeter scale. The pluton is interpanelled with various sized screens of meta-sedimentary rock, probably intruded at mid-crustal depths. The internal structure of the pluton is layered or foliated.

Mahan and Bartley (2000) interpreted the fabrics in the McDoogle Pluton to represent a sheared dike complex, describing ductile fabrics that are consistent with deformation along the Sierra Crest shear zone. Under this model, repetitive injections of dikes separated the elongate screens hosted in the pluton. Alternatively, the screens may have once been more irregular in shape and were deformed into parallel alignment with the fabric defined by the stretched mafic inclusions. Subsequently, there was a change in intrusion mechanics between the formation of the McDoogle Pluton and the Lamarck Granodiorite. McDoogle pluton subdivided numerous elongate screens of biotite schist whereas the Lamarck Granodiorite intruded in more of the cookie-cutter style, truncating the earlier formed screens.

Along the west side of Twin Lakes, the trail goes over a covered contact with the Twin Lakes Pluton (Jtl), and then traverses over a screen of biotite schist (Pzbs). The trail heads back into McDoogle Pluton (Kmd) for <100 m (<1/8 of a mile), and again over another screen of biotite schist.

The trail follows the drainage up 800 m in the Lamarck Granodiorite (Kl), traverses over a patch of talus (Qt), and heads slightly west to reach a contact with a thin, northwest-striking screen of biotite schist (Pzbs). In the valley bottom, the trail crosses an embayment in map pattern (near the drainage) filled by the Lamarck Granodiorite (Kl) containing an inclusion of Kmd. An embayment is a description of a contact as seen in one plane of exposure. This type of contact is similar in form to a shoreline defining a bay along a coast, in other words, it is very concave. The Lamarck Granodiorite is part of the John Muir Intrusive Suite, which is composed of the following plutons: Round Valley Peak Granodiorite, the Lake Edison Granodiorite, the Lamarck Granodiorite, and the Evolution Basin Granite (Fig. 65).

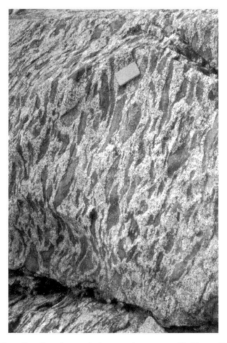

Figure 64. Diorite inclusion-rich outcrop of the McDoogle Pluton exposed near Pinchot Pass. The aligned inclusions of diorite define a marked foliation, and the lens shapes of the inclusions suggest that they were flattened when very hot, and possibly while partially molten. Notebook for scale is 19-centimeters long.

Lamarck Granodiorite (Kl)

The Lamarck Granodiorite, Maps 10-16 and 18-24, 89.6 Ma (Stern and others, 1981), 91.9 Ma (Coleman and others, 1995, preferred date), is a medium-grained, equigranular, generally seriate texture, hornblende-biotite-granodiorite. It has interstitial anhedral quartz and microcline grains. Plagioclase ranges from An_{52} to An_{12} (Frost and Mahood, 1987). $An_{\#}$ represents the percent of calcium versus sodium in a plagioclase crystal. Plagioclase having 90 to 100 % calcium (An_{90-100}) is called anorthosite. The most distinguishing feature of this rock are the large black hornblende crystals. The pluton is a large, northwest elongate, intrusive body, and is the oldest intrusion of the John Muir Intrusive Suite (Fig. 65). The foliations and joints in the pluton also strike to the northwest. The Lamarck Granodiorite contains numerous mafic inclusions and large internal diorite intrusions, interpreted by Frost and Mahood (1987) as resulting from magma mixing between the granodiorite and more primitive mafic melts. The eastern margin of the pluton cuts Jurassic plutons and metamorphosed Paleozoic sedimentary rocks.

Another feature of plutons yet to be resolved is exactly what percentage of the pluton was molten at any one time? Was the whole pluton one giant magma chamber filled by a melt or was there only a small magma chamber continually crystallizing material to its walls, gradually building up the entire pluton body. The presence of abundant mafic inclusions and small diorite intrusions within the Lamarck Granodiorite certainly suggests two melts of different compositions were being mixed during the formation of the pluton. This process of mixing seems to imply there was a large magma chamber into which two melts interacted.

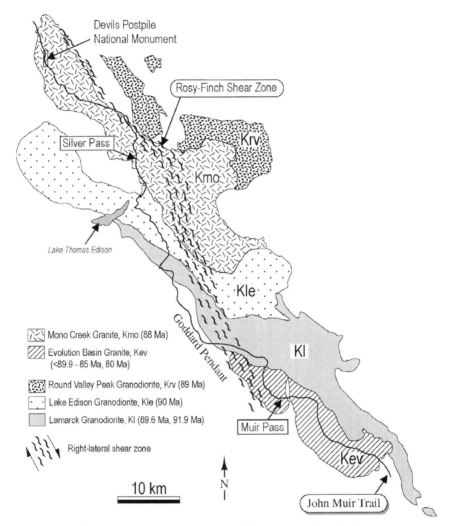

Figure 65. Geologic map of the John Muir Intrusive Suite.

Above the 3,440 m (11,287') elevation the trail is on the McDoogle Pluton (Kmd). The trail climbs numerous switchbacks on the south side of the ridge heading to Pinchot Pass (3,697 m, 12,130'). About 1/3 of the way up to the pass, the trail crosses a thin, northwest-striking screen of pelitic hornfels and quartzite (Pzph) (3,500 m, 11,484' elevation). The trail then returns to the Kmd unit. Numerous, thin slivers of meta-sedimentary rock are scattered throughout the pluton, but most are too small to show on the map (Fig. 54).

Pinchot Pass (3,697 m, 12,130', Map 10, waypoint 374300E, 4088516N)

Pinchot Pass is composed of the McDoogle Pluton, an inclusion-rich igneous rock (Fig. 64). Most of the inclusions were flattened parallel to the elongation direction of the metamorphic screens (Fig. 54). From Pinchot Pass northwards to Devils Postpile, the JMT traverses the entire length of the John Muir Intrusive Suite. To the west of the pass is Crater Mountain (3,923.8 m, 12,874') and to the east is Mt. Wynne (4,016.8 m, 13,179'). Mt. Wynne is made of Kmd containing abundant northwest-striking meta-sedimentary rock screens. Crater Mountain has similar screens engulfed by mafic plutonic rock.

Looking north into the South Fork of the Kings River, the first large lake is called Lake Marjorie. It lies within a glacial depression, or scoured low in the valley floor. The small peak to the east is composed of Kmd with several sills of Kl. To the south, numerous tarns lining the valley floor are in lighter colored rock formed by a thick sill of the Lamarck Granodiorite. Reddish rocks on the west valley wall are part of the Pinchot pendant, continuing southward, but narrowing to form the screen on the north face of Mount Cedric Wright.

Pinchot Pass is glaciated, having both polished and striated surfaces gently dipping to the north. To the east of the trail, at the low point in the pass, are granitic boulder erratics that did not originate from the McDoogle Pluton. These features clearly demonstrate ice was once higher than the pass- imagine the canyons both north and south completely filled by ice, carving away the cirques and horns and the pass that you are standing on!

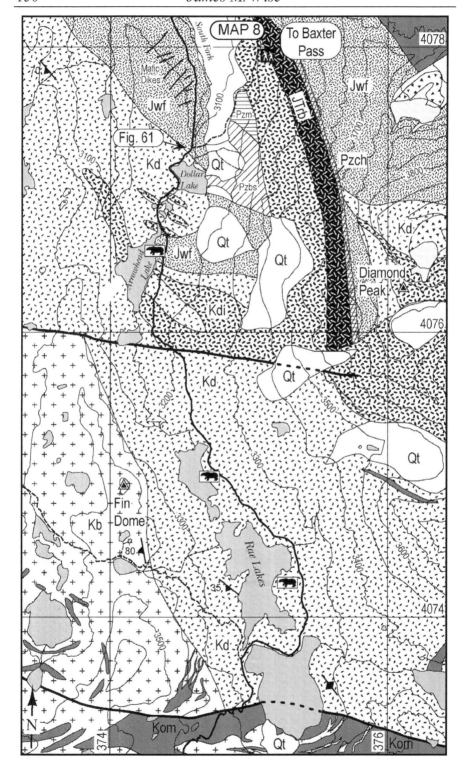

Geology of the John Muir trail 151

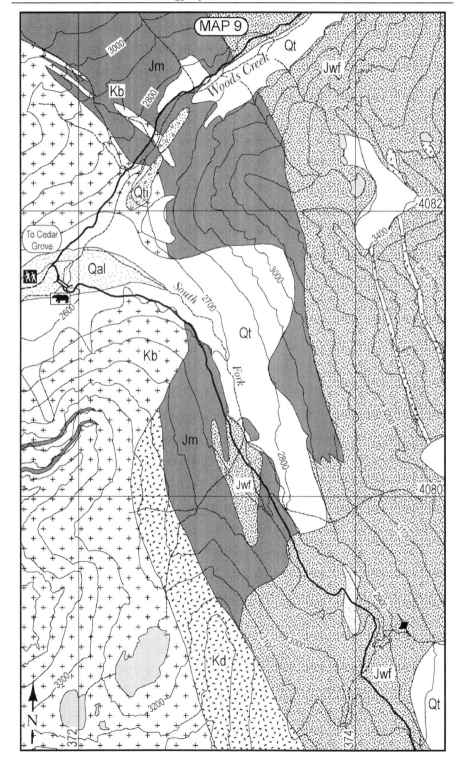

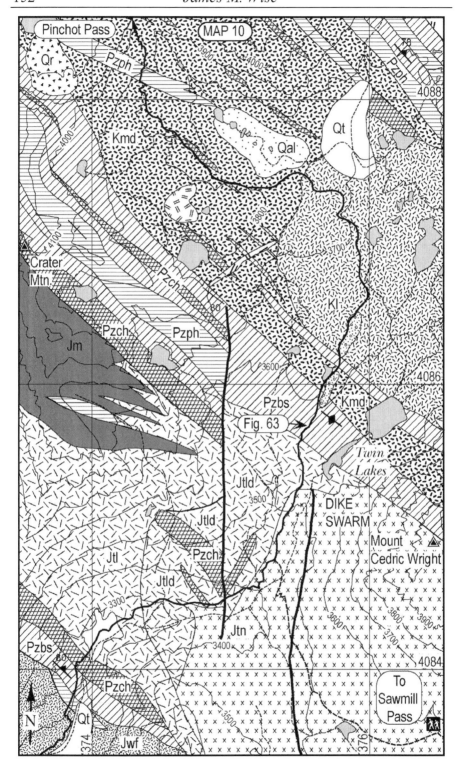

CHAPTER 5

PINCHOT PASS TO MATHER PASS

Access: Woods Creek, South Fork of the Kings River (cross country), and Taboose Pass
Distance: 16.7 km (10.4 miles).
Maps: 11-13.

This segment of the guide, although short, covers one of the more remote portions of the John Muir Trail. From Pinchot to Mather Pass, the trail crosses several plutonic units, and is in the watershed for the South Fork of the Kings River. The canyons are remote, the cirques raw with rock exposures, and the lower elevations shrouded in moraines and conifers.

Near the north side of Pinchot Pass, the trail crosses the contact between the McDoogle Pluton (Kmd) and the Lamarck Granodiorite (Kl) at the base of the main switchbacks; there is a marked color contrast between the rocks on either side of the contact. The trail descends over Kl and then nicks the side of sheared McDoogle Pluton. The sheared granodiorite is strongly foliated and oxidizing dark red. These outcrops just west of the trail contain a NW-striking mylonitic texture (ductile deformation) and were probably deformed in the Cretaceous Sierra Crest shear zone. The JMT descends to cross a stream and circumnavigates around the east side of Lake Marjorie.

The north arete of Mount Pinchot contains a large-scale fold pair exposed in the west-facing cliffs. The fold was cut by an EW-striking fault, and a prominent sill of Lamarck Granodiorite intrudes both the fold and the fault (Fig. 66). Fold deformation of the pendant here must be older than the Cretaceous, and probably happened in the Jurassic.

Halfway around the lake is the contact between Kl and the Cartridge Pass Granodiorite (Kcr). The contact is covered by dirt near the trail, and it is poorly exposed west of the trail. The trail descends and halfway around the next small lake it recrosses the contact between Kcr and Kl. The northern contact is better exposed to the east of the JMT (Fig. 67, Map 11). Kl is slightly darker than Kcr, and coarser grained. Both Kl and Kcr are foliated, however, the foliation is subtle because of the finer-grain size. The intrusive contact is sharp, steeply dipping northward, and has minor irregular curves.

The trail crosses two creeks, descends, and continues over a patch of talus (Qt) lining the valley floor. A lens of biotite schist (Pzbs), north of Lake Marjorie and west of the JMT, is extensively interlayered with granodiorite dikes and sills, which were tightly folded (Fig. 68). Both the folded granodiorite and well developed foliation in the biotite schist indicate contractional deformation, suggesting these screens did not form by separation along dikes, which are brittle-tensional features.

The trail traverses around the west side of a lake, comes to a trail junction that leads west to Bench Lake, and then farther north reaches the junction for **Taboose Pass and trailhead (ENTRY POINT, Map 11, waypoint 371986E, 4091435N)**. The trail descends from the junction, passing over the Lamarck Granodiorite (Kl) while going down numerous switchbacks to the South Fork of the Kings River. The trail fords the river, and ascends along its west side a short distance then fords a creek. Here an unmarked cross-country route heads west down the South Fork of the Kings River (waypoint 371529E, 4092282N). Next, the trail ascends to cross an E-W elongate hornblende diorite body (Km). The diorite is dark, easily distinguishable in the trail and surrounding slabs. The unit can be seen extending eastward all the way up the ridge. The southern contact (waypoint 371315E, 4093998N) is more obvious than the northern one. The diorite is heterogeneous, containing quartz veins, aplite dikes, darker inclusions, and changes in grain size (Fig. 69). Overall, it has the appearance of a migmatite, a rock that was partially melted. It is possible that the unit was originally a metamorphic rock and not a diorite. It is intertongued with the Lamarck Granodiorite along embayed, sharp contacts- both steep and low angle. At the 3,270 m (10,728') elevation the trail reaches a contact between Kl and Kcr (waypoint 370976E, 4095283N).

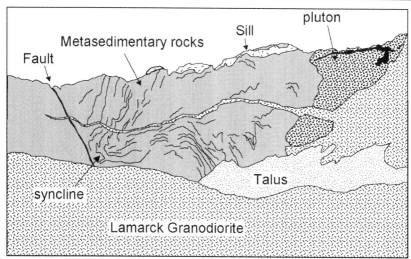

Figure 66. Field sketch looking northeast at the Pinchot cirque where folded metamorphic rocks are in a screen. The fold hinge lines are subhorizontal, and were cut by a EW-striking fault. In turn, a sill of Lamarck Granodiorite cuts the fold and fault. Thin lines show the traces of bedding.

Figure 67. Sharp intrusive contact between the Lamarck Granodiorite, the darker unit on the left side of the photo, and the Cartridge Pass Pluton, on the right side of the picture. The mechanical pencil is lying on the contact. Note the contrast in grain size across the contact. Also note that the foliation in the Lamarck Granodiorite is truncated at the contact.

Figure 68. Outcrop photograph of folded granodiorite and aplite sills in a screen between the Lamarck and Cartridge Pass plutons. Mechanical pencil for scale. The granodiorite appears similar to that of the McDoogle Pluton, but has yet to be dated. Folds deform foliation in the biotite schist and produced a second hinge surface cleavage in the intrusive rocks.

Figure 69. Photograph of highly foliated rock in a screen-like body. Unit contains thin stringers of granitic looking veins, indicating this rock was near to being melted, technically this rock may be called a migmatite (a rock that underwent partial melting). The unit was originally mapped as being a diorite, except it may indeed be a metamorphic rock. Lens cap for scale.

Cartridge Pass Granodiorite (Kcr)

The Cartridge Pass Granodiorite, Maps 11-12, 81 Ma (Evernden and Kistler, 1970), is a zoned pluton in which quartz and potassium feldspar increase towards the center. The potassium feldspars also increase in size towards the center. Mafic inclusions are more abundant near the margins of the pluton. The average mode is 23.4% quartz, 18.3% potassium feldspar, 47.8% plagioclase, 6.7% biotite, 1.7% hornblende, and 1.3% accessories (Moore, 1963).

Ascending into Upper Basin, the trail weaves back and forth over the covered contact between Lamarck Granodiorite (Kl) and Cartridge Pass Granodiorite (Kcr) (both appear similar from a distance), until it climbs along a medial moraine of Tioga till (Qg). The trail goes over Kl, talus (Qt), over a small, dark gray blob of diorite (Kd), best exposed to the north of an unnamed lake, and then over Kl. It climbs up material made of talus (Qt), and at the upper switchbacks it crosses the Kl/Kcr contact several times (Fig. 70). The contact is sharp with Kl being slightly darker gray.

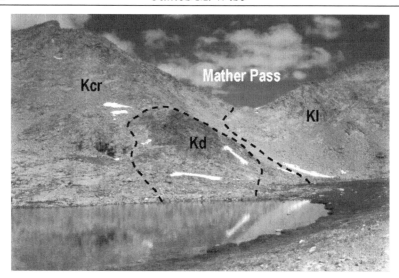

Figure 70. View northward towards Mather Pass. The dark rock behind the lake in the foreground is a diorite body (Kd). The Lamarck Granodiorite (Kl) is the lighter gray granitic rock composing Mather Pass and the ridge to the right. The slightly dark gray unit (not as dark as the Kd diorite), making the ridge to the left of the pass, is the Cartridge Pass Pluton (Kcr). The JMT crosses the sharp contact between these two units several times as it approaches the pass.

Mather Pass (3,688 m, 12,100', Map 13, waypoint 370206E, 4099138N)

The view to the south across the tarns of Upper Basin has a tree lined canyon of the South Fork of the Kings River in the distance. The distinct reddish-hued rocks along the skyline are part of the Mount Pinchot pendant meta-sedimentary rocks. Pinchot Pass lies in the saddle to the east of the reddish peak. The view to the north overlooks Palisade Lakes and the Palisade Crest composed of darker colored Inconsolable Granodiorite. The contact with the lighter Kl is sharp and exposed along the cliff sides. The left-side of Palisade Valley is marked by a dark patch of rock along the skyline composed of diorite (Kd). To the north is the Palisades Lakes, whose drainage joins the Middle Fork of the Kings River. Mather Pass is composed of the Cartridge Pass and Lamarck Granodiorites (Fig. 70).

Geology of the John Muir trail 159

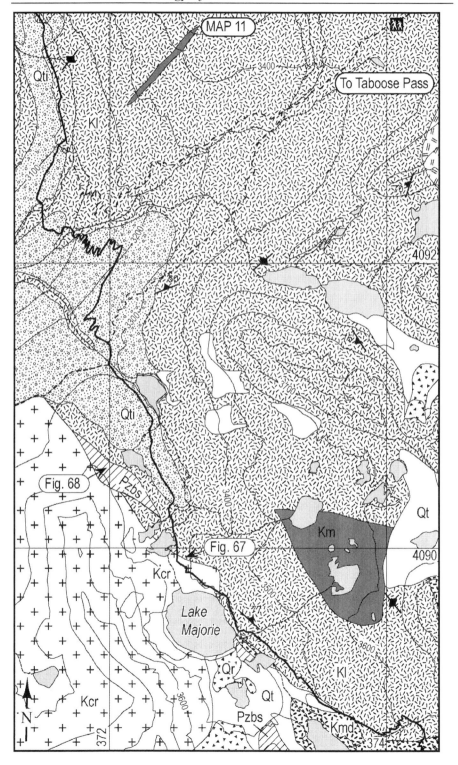

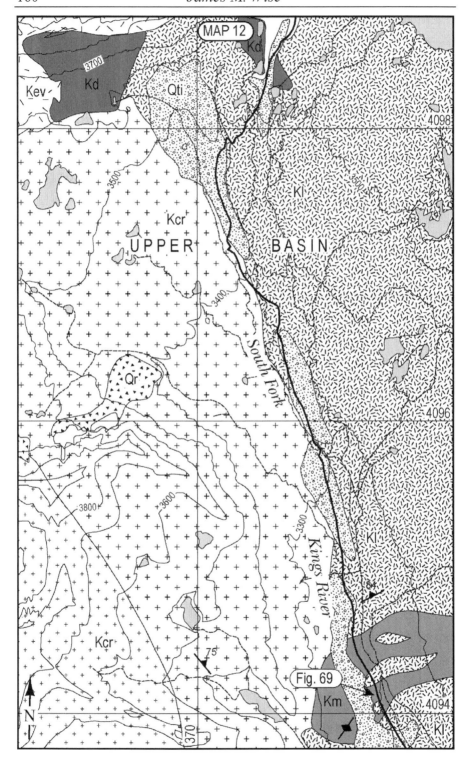

Geology of the John Muir trail 161

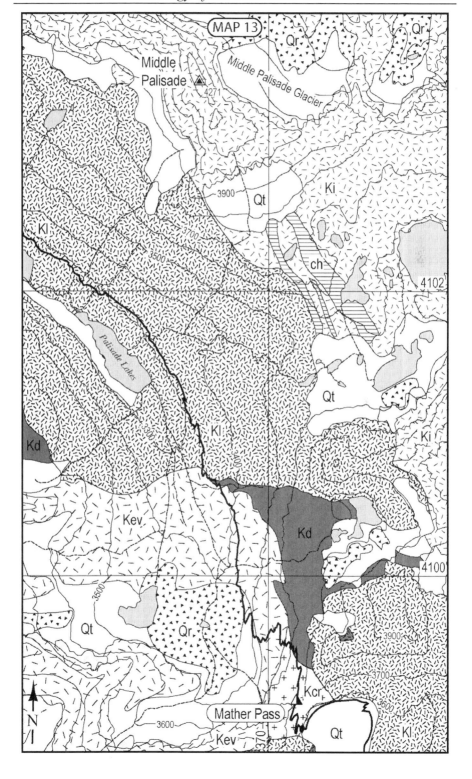

CHAPTER 6

MATHER PASS TO MUIR PASS

Access: Bishop Pass, Middle Fork of Kings Rivers (long approach)
Distance: 35.4 km (22 miles).
Maps: 13-17.

In the Mather to Muir Pass segment, the trail traverses over part of the Cartridge Pass Granodiorite, the Lamarck Granodiorite, the Evolution Basin Granite, part of the Goddard pendant, and crosses the watershed to the Middle Fork of the Kings River. The igneous rocks are part of the John Muir Intrusive Suite. The Goddard pendant contains metamorphosed volcanic rocks that are part of a regional Jurassic volcanic arc. Additional remnants of this chain of volcanoes are also seen along the JMT farther to the north near Devils Postpile.

Mather Pass (3,688 m, 12,100') is composed of the Cartridge Pass and Lamarck Granodiorites. The descent on the north side of Mather Pass encounters a complex section of mixed diorite and granodiorite. Large inclusions of diorite (Kd), some greater than 5 m in diameter, are angular shaped and frozen into the surrounding younger rock. Kd also includes numerous thin aplite dikes. The contact zone is characterized by complex interaction of different intrusives of probable different states of solidification during formation. The upper north-facing slopes near the pass commonly holds snow patches that can be very icy in the early morning, requiring caution. These snow patches early in the season may obscure or hide some of the above described rock units.

About halfway way down the slope (waypoint 370123E, 4099523N) the trail crosses the near vertical and sharp contact between the Cartridge Pass Granodiorite and the Evolution Basin Granite (Kev).

> **Evolution Basin Granite (Kev)**
> The Evolution Basin Granite, Maps 12-19, 80 Ma? (Dodge and Moore, 1968; Evernden and Kistler, 1970), is 20 km (12.4 miles) long and 5 km (3.1 miles) wide (Fig. 65). It is

> medium-grained, equigranular, quartz rich, and contains minor biotite. Overall, the rock has a slight pinkish hue, and is similar in appearance to the Bullfrog Pluton along the southern portion of the JMT. Near the fault contact with the Lamarck Granodiorite on the northern intersection with the JMT, the Evolution Basin Granite includes a fair number of irregularly shaped mafic inclusions. Evernden and Kistler (1970) reported dates from K-Ar analysis of biotite ranging from 79 to 85 Ma. North of Palisade Creek, Kev cuts Kl and has numerous angular inclusions of Kl near the contact. This style of intrusion suggests a stoping mechanism for the emplacement of the granite, because blocks of the older rock dropped into the younger magma chamber.

Next the trail skirts the east side of some talus (Qt), then descends the U-shaped valley, and re-crosses the contact between Kev and Kl at the 3,340.5 m (10,960') elevation near a stream (waypoint 369524E, 4100751N). The sharp, yet irregular, contact is exposed 20 feet below the JMT in glacial slabs. Thin aplite dikes cut the contact. Overall, Kev has fewer mafic inclusions. Marked color contrast between the two units helps in locating the contact. Along the contact is a diorite body (Kd) that pinches out laterally near the valley floor and several thicker lenses of diorite are also high on the both valley walls. The Evolution Basin Granite (Kev) truncates foliation in the Lamarck Granodiorite (Kl) at a 15 degrees, indicating Kev is younger. The contact also is exposed at an elbow of a small switchback in the JMT; the contact is sharp, striking 302 degrees, and dipping 86 degrees NE. Higher up and southward, the JMT crosses a 50-foot thick lens of Kd along the Kev/Kl contact, it is darker gray than Kl.

The trail circumnavigates Palisade Lakes along the northeast side. To the northeast is the Palisade Crest, Middle Palisade Peak (4,279.2 m, 14,040') and Disappointment Peak (4,241.7 m, 13,917') are all composed of the Inconsolable Granodiorite (a darker unit not crossed by the JMT). To the east, the north face of Mount Bolton Brown is ribboned by black dikes, which are part of the Independence Dike Swarm, bracketing the Inconsolable Granodiorite as being older than 148 Ma. The highly contrasting contact with Kl lies to the west of the crest.

Aplite sills are also abundant in the south valley wall of Lower Palisade Lake. The outlet area to Lower Palisade Lake was glacially scoured. From the outlet, the Goddard pendant and Devils Craig is visible to the west.

From Palisade Lake, the trail descends along the north side of Palisade Creek, crossing over Lamarck Granodiorite (Kl), a small diorite intrusion (Kd), a small body of Evolution Basin Granite (Kev), and then over the Lamarck Granodiorite. The diorite is hornblende rich, heterogeneous, and cut by aplite sills (Fig. 71). The very coarse-grained hornblende may indicate high water content in the magma. This has the effect of decreasing the rate of crystal nucleation and enlarging the grain size. The mafic unit west of the lake outlet is cut by laterally continuous aplite-sills, some of which are composite dikes containing mafic cores (Fig. 72). These dikes are important because they show two distinct compositions of magma coexisted in the same source magma chamber.

For the next 2 km (1 1/4 miles) the trail is on the Lamarck Granodiorite (Kl) then it enters Evolution Basin Granite (Kev), descending the numerous switchbacks of the Golden Staircase. The Golden Staircase, built in 1938, was the last major section of the John Muir trail to be constructed. This section of trail is usually a challenge to the south-bound hiker. Near the top of the Golden Staircase, a diorite block, engulfed by Kev, contains a high percentage of hornblende (crystals up to 2" in length), and abundant sphene. If you walk out onto the slabs of the rock south of the trail and look down towards Palisade Creek, numerous 5' to 35' diameter blocks of similar diorite are cleanly exposed in the glaciated slabs, delimiting a dark gray and white mosaic, or distorted triangular checkerboard. Far up on the south valley wall in the U-shaped hanging valley is an active rock glacier built of large blocks perched and ready for gravity to nudge them over the edge, crashing down to the large talus cone below. Between the Golden Staircase and Glacier Creek, the JMT crosses glacial slabs adjacent to Palisade Creek. These exposures have broken pieces of Kl engulfed by Kev aplite. The north valley wall exposes multiple blocks of Kl at different degrees of disaggregation by Kev along the Kev/Kl contact, indicating Kl was brittle during intrusion of Kev, and suggestive of stoping of Kl into Kev (Fig. 73). As the trail wanders over the Lamarck Granodiorite (Kl) foliations in the pluton strike NW and dip 70 degrees to the northeast.

The trail comes to a junction at Deer Meadows; the trail to the south leads to a dead end. The south fork was built to access Cataract Creek for a mining prospect (this old trail is overgrown and difficult to find). North of Deer Meadows is North Palisade (4,340.7 m, 14,242'). Next the trail stays on the north side and descends along Palisade Creek over talus (Qt) for 3.2 km (2 miles). The view westward down Palisade Creek is dominated by Devils Craig piercing the skyline, a collection of horns and cirques. The darker rock is composed meta-volcanic rocks of Goddard pendant, sharply contrasting with the lighter colored Evolution Basin Granite. The trail comes to a junction for the Middle Fork of the Kings River trail (waypoint 359784E, 4101689N). This west-bound trail makes a lengthy trip to pass Tehipite Dome, climbs northward out of the Middle Fork, and then intersects a network of trails southeast of Wishon Reservoir in the Rough Spur quadrangle. Beneath the talus, the trail descended over covered Kl, diorite (KJd), and Kev. At this confluence of two main canyons, Palisade and LeConte, it is worth considering their form and the processes that developed the canyons.

Figure 71. Photograph of a thin aplite sill cutting coarse-grained diorite. The large black crystals are pegmatitic hornblende. This diorite body has some similar textures to that exposed in the Onion Valley layered mafic complex outcropping at Glen Pass and Painted Lady. Lens cap for scale.

Figure 72. Composite sill of mafic rock in the center enveloped by lighter colored aplite. The sill dips nearly 30 degrees and is cutting the steeply foliated Lamarck Granodiorite. Note how the darker portion of the dike is in elongate, irregular ellipses or pods. This texture is suggestive of both compositions of rock being molten at the same time when injected into the sill. Lens cap for scale.

Figure 73. A large mafic inclusion deformed by brittle offsets across aplite sills (the upper inclusion had a displacement to the left). Note the angularity of the inclusion contacts with the surrounding granodiorite. Outcrop lies north of the John Muir Trail, down hill from the Golden Staircase. These mafic blocks may be remnants of a diorite dike intruded into the host of granodiorite or may be inclusions derived from the Lamarck Granodiorite.

Development of the U-shaped Glacier Canyon Profile

The comparison of river eroded canyons versus those of glaciated country long ago led to the observation that they form shapes described as V and U in profile view, respectively. Figure 74 shows a location map for several valley profiles across LeConte, Palisade, and Middle Fork of the Kings River Canyons. Both the LeConte and Palisade Canyons have the characteristic U-shape profiles of glaciated valleys. The Middle Fork of the Kings River collected the combined flow of these glaciers, and therefore one might expect deeper incision of the canyon, or perhaps more perfectly developed U-shape profile. John Muir used the argument of several glaciers combining to form a trunk glacier provides the focus to form large canyons such as Yosemite Valley or Kings Canyon. The greater depth of the main Yosemite Valley as compared to the classic perched hanging valley at Bridalveil Falls is also attributed to the relative size of glaciers. Profile G in Figure 74 does not exactly fit this pattern, showing more of a V-shape than the next profile located downstream.

Not all glaciated canyons obtain the ideal U-shape, for example, Tenaya Canyon at Yosemite Valley was certainly glaciated numerous times, and still retains more of a V-shape profile. Guyton (1998) suggested that this reflects the massive bedrock character over which the glaciers flowed. Likewise, the glaciated Grand Canyon of the Tuolumne River has a V-shaped profile, which Huber (1981b) related as glacial plucking following pre-existing exfoliation sheets that parallel the canyon walls. Rock masses lacking joints are less easily modified by a glacier from the initial V-shaped developed by stream incision. Integral in evaluating the form of canyons, whether stream or glacial in origin, is the interplay of the bedrock geology, a complex study in itself that is the emphasis of the branch of geology called geomorphology and beyond a complete review here. Nonetheless, the lower profiles drawn in the Middle Fork may have an influence from the bedrock where the river crosses the rocks of the Goddard roof pendant. All the profiles shown in Figure 74 may have some modification by talus along the flanks of the canyons. The apparent more pronounced V-shape in profile G may also be explained by development talus deposits.

How do glaciers carve out a U-shape valley profile? First, it is generally assumed that a V-shape profile preceded the glacier occupying the valley, an assumption that may not always be true. Streams cut into rocks by focused erosion at the base of the stream channel, another complex topic involving several variables such as river discharge, channel slope, and the amount of sediment carried by the stream. Streams lower or incise their channels while at the same time additional material is carried down the valley walls by a completely different set of physical processes, such as tributary streams, slope wash, creep, and mass wasting; all these are driven by gravity. The additional material placed in the river may enhance the stream erosion or overwhelm the system, filling the channel with more sediment than the river can carry and temporally halt downward erosion. Finally, the V-shape can have variable slope angles depending on the rate of down cutting by the stream, strength of the bedrock, and the rate of mountain uplift and/or lowering of the base level.

Glaciers have several important differences from streams. First, the erosion and transport agent is a solid that flows in a completely different way than running water. Second, glaciers can accumulate substantial thickness and exert much higher pressures than flowing streams. Third, the physical mechanism by which the bedrock is eroded through **abrasion** and a process called **plucking** is different. Rivers abrade their channels by the impacts of the suspended particles in the flow and by rolling cobbles and boulders. The amount of pressure, or stress, caused by these impacts is limited, however, the frequency of the abrasion is high. On the other hand, glacial abrasion of particles imbedded in the ice can be at high pressures. The key factor in the pressures generated by glaciers is not just the overlying weight of the ice; it is the lateral motion of the glacier that imparts a shear stress. This grinding action produces large quantities of rock flour that is eventually released and washed down with stream flow. The second process of bedrock erosion by glaciers is plucking, which is a variation of frost wedging where water seeps into cracks, expands, and further fractures the rock. Once the bedrock rock is broken by ice expansion, the loosed block may be held by the glacial ice and removed, or plucked out, as the glacier moves down valley.

These two processes, abrasion and plucking, may not seem overly complicated, but glaciers have some additional characteristics that make their study interesting. For plucking to work, water must be occasionally present, probably as a thin film between the ice and the rock. If the glacier is very cold and no water is formed, then plucking will probably not be involved in erosion. Some active glaciers change character through time so that their velocity fluctuates. When glaciers move relatively slow and then undergo sudden bursts in speed, they are called surging glaciers. One explanation for the increased speed of the glacier is the formation of a layer of water at the glacier's base that acts as a lubricant or, if thicker, may suspend the glacier. Surging glaciers are perhaps ideal for plucking if their motion represents a freeze-thaw process. Most glaciers have streams issuing from their terminus and therefore river erosion beneath the glacier may be an important part of the system. Glaciers, like rivers, also transport all the valley wall detritus carried down by gravity. Much of this material falls upon the top of the glacier, or is drug along its side in the lateral moraines, and not all of this material is available to act as tools in abrading the bedrock.

Figure 74 shows two main schools of thought on the relative degree of glacial erosion of an initial V-shape profile. One idea, Model 1 in Fig. 74, maintains that glaciers are poor incisors, unable to cut downward. The valley is modified from a V to U shape only by lateral erosion of the valley walls, so the characteristic U-shape represent widening, not deepening of a valley. Right at the base of the V a glacier has reduced flow because of friction and this results in what is called a dead zone of lower erosion. Erosion to either side of the dead zone is increased because of the higher velocity of ice flow. Although, if subglacial stream flow happens the base of the canyon may continue to be lowered. Evidence for subglacial flow are potholes eroded into the bedrock. The other possibility, Model 2 in Figure 74, is that glaciers do erode downward and do so in equilibrium form which approximates a cylinder. Note that one does not exclude the other, there are probably numerous cases where both of the above scenarios were important, and it is likely that model 1 proceeds model 2. Numeric modeling by Harbor (1992) suggests that glaciers will erode a valley to a U-shape regardless of the initial valley shape. One final example is worth mentioning on the capacity or ability of a glacier to erode downward. Along the JMT are numerous lakes and tarns, situated in bedrock basins, which clearly cannot have been formed by stream erosion. These lakes are in lows excavated by glaciers.

One last comment on Figure 74, some of the profiles have several different valley wall slopes that may seem similar to the cartoon presented showing a glacier modifying a V-shaped valley. All the profiles drawn in Figure 74 were located along the divides to either side of the canyon. These divides are glacial controlled aretes from tributary glaciers feeding the main canyon. The changes of slopes may reflect the trunk canyon incising deeper than the tributary canyons, thereby explaining the multiple slopes.

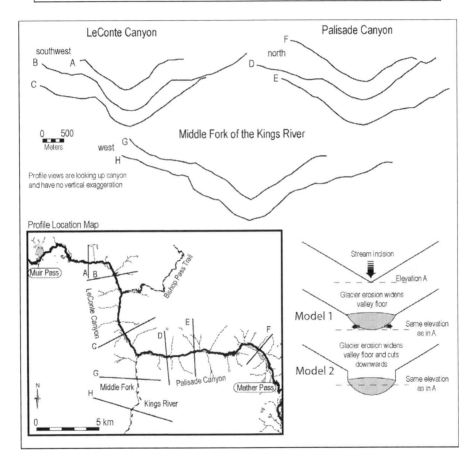

Figure 74. Canyon profiles and their formation by stream and glacier incision.

From the confluence of the two canyons, the trail continues northward and up LeConte Canyon. The canyon was named after Joseph Nisbet LeConte, once a professor of engineering mechanics at the University of California, who explored the canyon in 1908 with James Hutchinson and Duncan McDuffie. LeConte's father was

Joseph LeConte, the first geology professor at the University of California and one of the early geologists to study the Sierra Nevada in the 1870's. The glacially carved, U-shaped valley heads northwest and is lined by talus (Qt) and alluvium (Qal) for 2.4 km (1 2 miles) along the east side of Grouse Meadow. The river at Grouse Meadow forms well-developed meandering curves similar in geometry as those much larger rivers.

The prominent peak overlooking LeConte Canyon on the west side is an arete called the Citadel, and to either side are glacial carved U-shaped valleys. The steep canyon walls are riddled by several parallel sets of continuous joints and are actively exfoliating, forming aprons of talus at the base of the cliffs. The JMT crosses the Dusy Branch stream by a footbridge, crosses a talus slope, and then stays on the east side of the Middle Fork of the Kings River. The trail goes north at the junction for Dusy Branch near the LeConte Ranger Station; Dusy Branch leads east to Bishop Pass (3,648.9 m, 11,972', an arduous 6 mile climb from the JMT) and on to the South Lake trailhead which is reached from **Bishop (ENTRY POINT, Map 16, waypoint 358394E, 4106318N)**.

Langille Peak, to the west of Little Pete Meadow, dominates the canyon wall, rising in cliffs nearly as impressive as those of Yosemite Valley. In certain times of the day some of the glacier polished walls reflect like a vast mirrors. The glaciers filling LeConte Canyon in the past must have been impressively thick, as estimated from the height of the valley walls and the numerous tributary valleys contributing ice to the main glacier.

At Little Pete Meadow the trail crosses a small section of cover (Qal). East of Little Pete Meadow, high on the valley wall, the contact between Kev and the darker Kl is clearly exposed. Once on the southeast side of the meadow the trail is back into Kev. The trail is built on more Qal, the material underlying Big Pete Meadow, and then crosses over some talus (Qt).

The ridge of dark rocks looming over the west side of the upper portion of the canyon is named Black Divide and it is composed of the meta-volcanic rocks of the Goddard pendant (Fig. 75 and 76). The trail makes steep switchbacks climbing out of LeConte Canyon, passing over the Evolution Basin Granite, staying on the north side of the Middle Fork of the Kings River, and passes a small lake on the south side.

The trail again ascends switchbacks, and then runs around the south side of a small clear lake. Climbing west of the lake and along a stream, the trail traverses the eastern contact between the north-pointing spur of the Goddard pendant and the Evolution Basin Granite; the contact is vertical and sharp. The exposure immediately north of the trail is rubbly. To the south, the outcrop across the stream better exposes the contact, but it is difficult to approach. These volcanic rocks of the Goddard pendant were formed at about 140-143 Ma (Saleeby and others, 1990). In general, the rocks are massive, very fine-grained, and colored dark gray, black, dark gray-green, and rusty orange to brown. Some of the lighter gray units are metamorphosed tuff. The meta-volcanic rocks were intruded by numerous dikes from the nearby Evolution Basin Granite, resulting a more complicated appearance.

The trail then continues in the metamorphosed volcanic rocks of the Goddard pendant, ascending a narrow rubble filled valley. The trail next crosses a thin N-S trending body of sheared fine-grained granite of the Goddard pendant (Jgg).

> **Sheared finer-grained granite of Goddard pendant (Jgg)** is older than 158 Ma (Chen and Moore, 1982). The deformation that sheared this rock may be related to the Late Cretaceous Sierra Crest shear zone system (Greene and Schweickert, 1995).

The trail leaves the JTrmv unit at the outlet of Helen Lake (Fig. 77), and looking to the north the metamorphic and igneous rocks contrast with one another dramatically. Near the western contact, the Evolution Basin Granite contains abundant fragments or xenoliths of metamorphosed volcanic rocks. The trail passes Helen Lake (3,540.7 m, 11,617') on the south side, fords the inlet to the lake, and then climbs westward (usually over snow fields) via switchbacks to gain Muir Pass.

Figure 75. Photograph looking northwest at the Goddard pendant exposed on Black Divide (darker gray rocks on the crest above the snow patches). Note the U-shaped valley profile in the center of the photo.

Geology of the John Muir trail 175

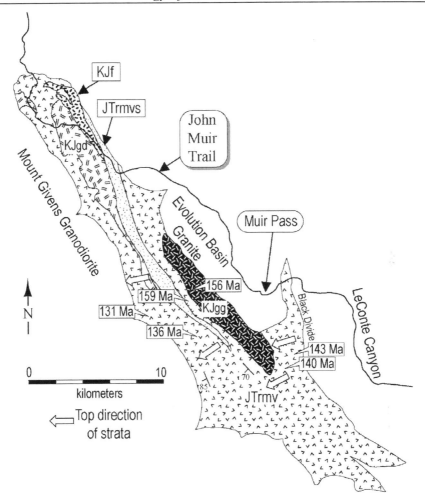

Figure 76. Generalized geologic map of the Goddard pendant. Note the cuspate or concave segments in the outline of the pendant- these are places intruded by separate granitic plutons, almost in a cookie-cutter style. Age data in boxes are from Tobisch and others (1986), geology generalized after Bateman and Moore (1965).

Figure 77. Photograph looking northward at Helen Lake. The dark rocks in the center of the photo are part of the Goddard pendant, exposed in a north trending spur forming a screen. Light gray rocks on the right side of the photograph are plutonic.

Muir Pass (3,646.8 m, 11,965', Map 17, waypoint 351643E, 4108406N)

To the southwest of Muir Pass is Mt. Solomons (3,972.5 m, 13,034'). Directly south of the pass, the dark rocks along the summit of Black Giant are part of the metamorphosed volcanic rocks of the Goddard pendant. The contact between the Goddard pendant and the lighter colored Evolution Basin Granite is clearly visible along the base of the ridge. The basins to the south are the start of LeConte Canyon through which the Middle Fork of the Kings River flows. From Muir Pass looking northward, is Wanda Lake, and the dark-colored rocks to the northwest of the lake comprise the summit of Mount McGee. Muir Pass formed the head to the South Fork Glacier of the San Joaquin River during the Tioga stage glaciation, and forms a divide between two major watersheds. Finally, Muir Pass is somewhat distinctive for the landmark it holds- a stone hut built to shelter hikers.

Geology of the John Muir trail

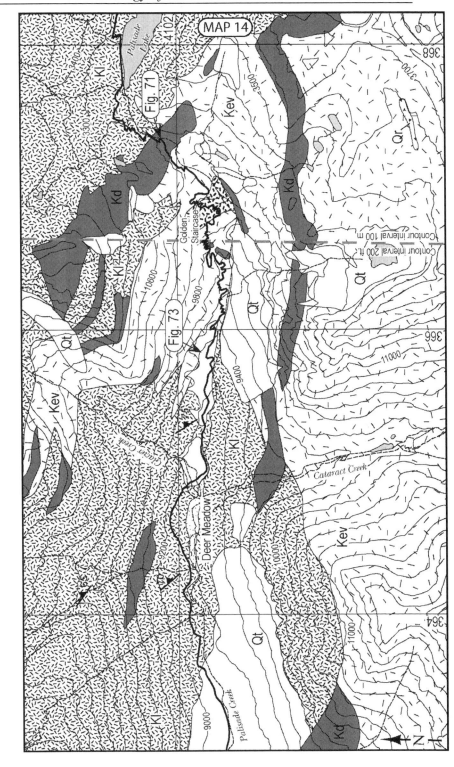

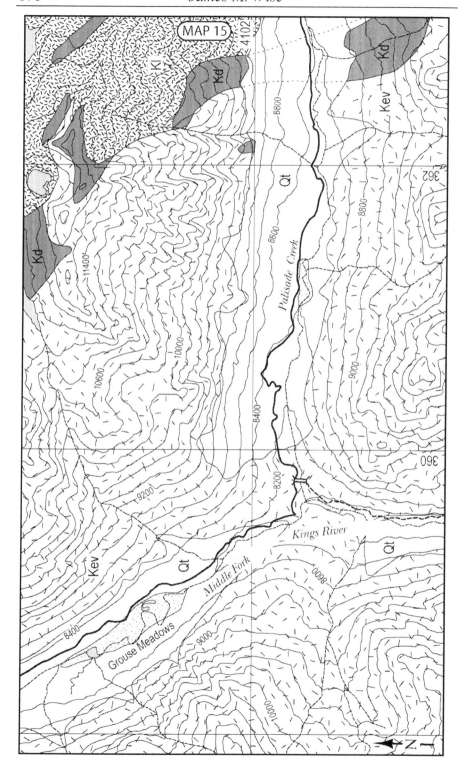

Geology of the John Muir trail

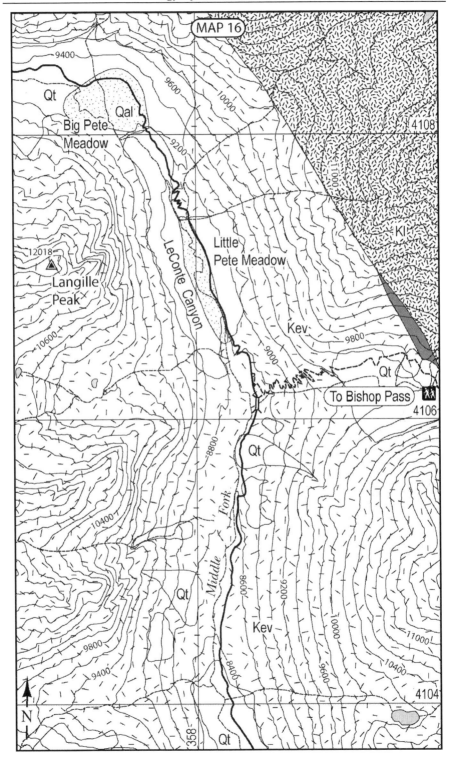

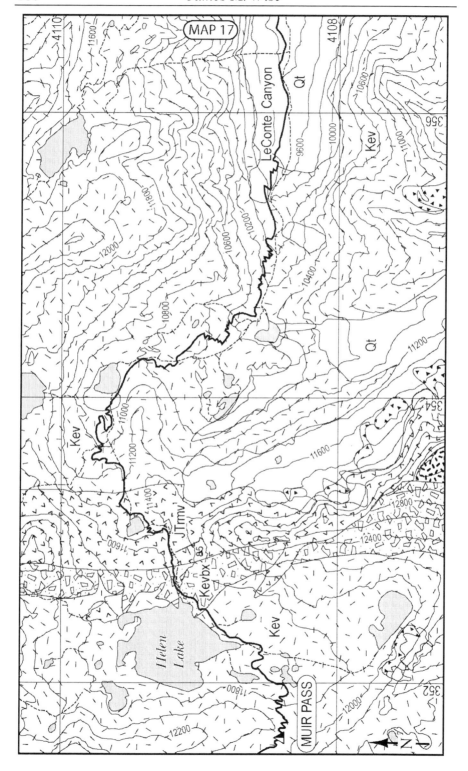

CHAPTER 7

MUIR PASS TO SILVER PASS

Access: Bishop Pass, Piute Pass, Lake Florence, Lake Thomas Edison, and Mono Pass.
Distance: 80.5 km (50 miles).
Maps: 17-27.

The Muir to Silver Pass segment of the John Muir Trail weaves in and out of rocks belonging to the John Muir Intrusive Suite and older metamorphosed volcanic rocks of the Goddard pendant plus sheared older plutons. Some of the lower elevations found along the entire trail are in this segment. Whereas most trail segments are divided by major subdividing passes, this part also includes Selden Pass and Bear Ridge, both significant climbs in themselves. This segment of the guide also has several access points from both the west and east sides of the Sierra Nevada.

The JMT descends northward from Muir Pass into a broad U-shaped valley to pass a tarn on the north side, and then around Lake McDermand on the southwest side. The north side of Muir Pass is commonly covered by a large snow field. The trail circles Wanda Lake (3,482.5 m, 11,426') on the east side. Across from the peninsula pointing eastward in Wanda Lake is the contact between Kev and Qg. Southwest of Wanda Lake is Mount Goddard (4,135.3 m, 13,568'), to the south is the Goddard Divide, and to the east is Mount Huxley (13,086') and Mount Warlow (4,025 m, 13,206'). The trail fords the outlet to Wanda Lake, descends high above the west side of a small lake, and then passes the intensely blue Sapphire Lake (3,342.3 m, 10,966') on the west side.

From Sapphire Lake, the trail descends along the west side of the creek to the contact between the Evolution Basin Granite (Kev) and the Lamarck Granodiorite (Kl). The contact is located several hundred meters north of the inlet to Evolution Lake, at the base of some short switchbacks over glacially carved slabs (waypoint 349770E, 4113657N). Immediately to the east of the contact where it intersects the trail, the stream takes a bend nearby a single small tree. Near the contact, the Evolution Basin Granite contains irregular mafic inclusions, whereas mafic inclusions are absent in the Lamarck Granodiorite near this part of the contact. The contact is a brittle fault oriented N59°E, 88° NW and has 10-cm thick layer of cataclasite along it. Cataclasite a fine-grained rock that was crushed and ground up by motion of a fault. Several minor displacement cataclasite zones branch off the main fault and head directly north. Most of the splay faults attenuate or die off to the north. Several of the conjugate faults (paired faults typically at a 60 degree angle from one another that formed simultaneously) moved with left-lateral and right-lateral offset where they cut aplite dikes. The fault contact continues to the east to run through the middle of Mount Spencer (3,788 m, 12,431').

Just before crossing the inlet to Evolution Lake (Map 19), examine the glacially carved slabs to the west. The Lamarck Granodiorite here is compositionally layered and does not have mafic inclusions. The biotite and hornblende in the rock are concentrated in 3 to 4 cm bands separated by 10-cm bands of lighter colored rock. This subtle layering dips 28 degrees to the northwest. The varying compositions reflect changes occurring in the magma chamber, such as intrusion of new mafic melt, or expansion of the chamber. Overlooking the trail to the east is the summit of Mt. Darwin, which may be a remnant of a Cenozoic erosion surface and part of the summit upland plateaus.

The trail fords the inlet stream to Evolution Lake via large stepping stones, and continues around the northeast side of Evolution Lake (3,307 m, 10,853'). To the east is Mt. Mendall (4,178.6 m, 13,710') and Mt. Darwin (4,215 m, 13,830'). Both of these craggy appearing, highly jointed mountains are composed of the Lamarck Granodiorite.

Surrounding Evolution Lake are beautiful glacially polished slabs. The trail descends numerous short switchbacks from Evolution Lake to drop into the upper portion of Evolution Valley. Foliations in the Lamarck Granodiorite are striking NW and dipping 30 to 70 degrees NE. The trail gradually drops to lush Colby Meadow, and then to McClure Meadow, staying on the north side of Evolution Creek (the mosquito population at these meadows can be substantial). McClure Meadow is named after the California State Engineer, Wilbur F. McClure, who assisted in constructing the John Muir Trail. Just before the ford to Evolution Creek, the Lamarck Granodiorite is exposed in various scattered near horizontal slabs. The rock is foliated, the fabric striking northwest and dipping to the northeast, and the mafic inclusions are stretched parallel with the foliation.

There are two fords to Evolution Creek; the westernmost ford is most often used by stock. The eastern ford, probably the better option for the backpacker, involves following a very faint path across the meadow to a wide, fairly easy, sandy crossing (Map 20, waypoint ~342097E, 4117926N). Once on the south side of Evolution Creek, follow the path downstream to the west ford.

From the western ford, the trail descends pass several large dark colored septa of meta-volcanic rocks contained in the Lamarck Granodiorite. The number of septa increases the closer to the contact with the east margin of the Goddard pendant.

From Evolution Meadow, the trail descends many switchbacks down 244 m (800') to a meadow at the confluence of Evolution Creek and South Fork of the San Joaquin River. Near the upper portion of the switchbacks, at 2,700 m (8,860') elevation, the trail intersects the igneous contact between the Lamarck Granodiorite and the Goddard pendant (Map 21, waypoint 341134E, 4117743N). The JMT diagonals across the Goddard pendant as it follows the South Fork of the San Joaquin River. The contact is best-exposed 5 m north of the trail in glacially carved slabs (Fig. 78). The contact is sharp, in places there is 1-cm zone of assimilation (melting) of the metamorphosed volcanic rocks. Looking to the north across the Evolution Valley, the near vertical contact is exposed in the cliff face (Fig. 79). The Goddard pendant composes the dark gray rocks and the Lamarck Granodiorite the lighter rocks to the east. The Lamarck Granodiorite is not very foliated near the contact, and the few mafic inclusions are relatively undeformed.

The trail descends down a steep ridge, using switchbacks blasted into the rock, and then near the base of the descent it crosses over meta-sedimentary tuff (JTrms). Meta-sedimentary tuff represents material that erupted violently, but redeposited in the sedimentary environment and metamorphosed from heat and pressure.

Figure 78. Intrusive contact (arrow) of the Lamarck Granodiorite (Kl) and the Goddard pendant in a glacial carved exposure a few meters north of the trail (Map 21). Lighter colored rock is the granodiorite. The darker colored metamorphic rock of the pendant has a well developed foliation subparallel to the contact.

Figure 79. Contact between the Lamarck Granodiorite (Kl) and the Goddard pendant (JTrv) exposed in the north canyon wall of lower Evolution Canyon. View to the west.

Goddard pendant

Volcanic bedding the Goddard pendant is parallel and steep, and generally has stratigraphic tops tilted towards the west. The volcanic rocks include felsic to intermediate tuffs, breccias, and lesser rhyolite and mafic lavas. Because the rotated beds are near vertical the present map view of the pendant provides a cross-section through the volcanic complex. Six U-Pb dates reported in Tobisch and others (1986) range from 131 to 159 Ma, placing this section of volcanic rocks into the Jurassic (Fig. 76). The steep dips are probably from fold deformation of the volcanic sections, which also accounts for the metamorphic grade and cleavage development of the rocks. The volcanic bedding strikes to the northwest and overall is parallel for the entire length of the pendant.

The volcanic rocks represent just a small part of the regional scale Jurassic continental margin that was actively being subducted by oceanic rocks to the west. The Jurassic chain of volcanoes ran from north of Oregon, along the Sierra Nevada, through the eastern Mojave desert area, and on into western Arizona (Busby-Spera, 1988; Dunne and Walker, 1993). The volcanic arc continued farther south into Mexico and subsequently part of it was displaced by the late Cenozoic San Andreas fault. The displaced portion of the arc includes the Santiago Peak volcanic rocks in coastal southern California and Baja California. To the east of the volcanoes was a large **back arc** area of arid climate that was covered by one of the world's larger sand dune fields, which is now frozen into the sandstones of the Aztec Formation. This formation, covering large parts of Utah and Arizona, contains fine-grained volcanic detritus that was transported eastward from the Jurassic volcanoes. Closer to the Sierra Nevada section of the arc, the back arc area was occupied by a deep marine basin that was filled by shales, a group of rocks informally referred to as the "Mud pile" in northwest Nevada.

The east side of the pendant has near vertical, planar, intrusive contact between both the Evolution Granite and the Lamarck Granodiorite. Along Evolution Valley's north wall, the contact dips about 85 degrees towards the southwest. Similarly, the western contact with the ~90 Ma Mount Givens

Pluton is also near vertical (Tobisch and others, 1993: Tobisch and others, 1995; McNulty and others, 2000). McNulty and others (2000) proposed that piston-style sinking of the magma chamber floor accommodated the space for the evolving Mount Givens Pluton, and thus the Goddard pendant is bounded on the west by a NW-elongate passive type magma chamber (Fig. 80). This model uses a series of north-south striking feeder dikes, beneath the magma chamber, that are oriented perpendicular to the main direction of tectonic tension. However, at greater depths in the crust, approximately deeper than 15 km, rocks undergoing strain behave as a plastic material or in what is called ductile deformation. In this environment rocks tend to flow instead of fracture, depending upon the strain rate. Therefore, the model of deep-seated feeder dikes may be inconsistent with the type of deformation expected under these temperature and pressure conditions. In support of the model, the wall rock adjacent to many plutons do have dikes and sills. Brittle features, such as dikes and faults, can form in rocks that are more likely to contain ductile features by increasing the strain rate (the speed at which the force is applied to the rock and at which the rock deforms). If the emplacement of the Mount Givens Pluton was controlled by deep-seated NW-striking faults these may have been part of the regional early Cretaceous right-lateral Mojave-Snow Lake fault.

The Goddard volcanic complex may be similar to the volcanic rocks at the Ritter Range and Mount Morrison pendants. Both the Ritter Range and Mount Morrison pendants have volcanic strata spanning from the late Triassic to the middle Jurassic. Moreover, the Ritter Range pendant also includes Cretaceous metamorphosed volcanic rocks. Taken together, the meta-volcanic rocks represent the volcanic cover erupted throughout the three main periods of granitic intrusion in the Sierra Nevada batholith.

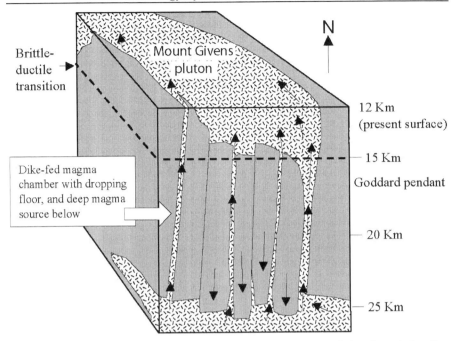

Figure 80. Schematic diagram showing magma injection into the Mount Givens pluton with provided by the floor dropping into a lower reservoir of melt (modified from McNulty and others, 2000).

The JMT passes over the South Fork of San Joaquin River by way of a wooden bridge. The river continues up to the south in Goddard Canyon, parallel to the elongate direction of the Goddard pendant. The trail fords an unnamed stream, skirts the west side of a meadow laden by fallen logs and trees, and then recrosses to the north side of the South Fork of the San Joaquin River by another wooden bridge. Several good exposures of meta-sedimentary tuff (JTrms) are along the trail, having gently dipping beds that are probably near the crest of a fold. The trail descends into Aspen Meadow (Map 21), going by barren, light-colored talus slopes of blocks composed of felsic igneous rocks (KJf).

From Aspen Meadow, the trail descends over talus and passes numerous discontinuous exposures of deformed granodiorite (KJgd), many of which were intruded by mafic dikes. In the narrow part of the canyon, large sections of the forests have been knocked over from snow avalanches coming down the south valley wall. The trail continues through a blasted section circling a knob, where the valley bottom becomes narrow. Just east of Piute Creek, an exposure in glacially carved slabs of granodiorite (KJgd) contains several mafic dikes that cross-cut one another. The dikes appear to be radiating out from a small mafic body that intruded the granodiorite. The granodiorite is locally deformed, indicating the rock is probably older than the Cretaceous John Muir Intrusive Suite to the east and the Late Cretaceous immense pluton to the west of the Goddard pendant called the Mount Givens Granodiorite.

Piute Creek marks the boundary between Kings Canyon National Park to the east and the John Muir Wilderness to the west. The JMT crosses Piute Creek by a bridge, and nearby are good exposures of the granodiorite (KJgd) intruded by mafic dikes. This granodiorite body is a large elongate intrusion into the meta-volcanic rocks of the Goddard pendant (Fig. 76). The granodiorite in places has small anastomosing (braided pattern) brittle faults that deformed the rock. At Piute Creek the granodiorite is foliated, and contains several small mafic inclusions. **(PIUTE PASS/CREEK ENTRY POINT, Map 22, waypoint 337427E, 4121211N).**

From Piute Creek the trail gradually descends over poorly exposed Tioga till, recent valley fill, and low rounded exposures of the granodiorite (KJgd). The trail takes the north fork at the junction for the **Florence Lake trail (ENTRY POINT, Map 22, waypoint 334934E, 4121350N)**. Food drops may be arranged at the Muir Trail pack station along the Florence Lake trail, about 2.5 km to the west. Also, along this trail, on the south side of the San Joaquin River at Shooting Star Meadow, there is a good hot spring (waypoint 333115E, 4122354N). Florence Lake, actually a large reservoir, has a ferry service to the trailhead on the west side of the lake. The reservoir was made in 1926 to store about 64,406 acre-feet of water for San Joaquin Valley irrigation.

From the trail junction, the JMT gradually contours up the north valley wall and comes to another trail junction for the Florence Lake trail, also called the short cut to Blayney Meadows. The trail climbs pass numerous poor exposures of undifferentiated meta-volcanic rocks, switchbacks up and over Qg and Qt covered slope vegetated with manzanita and sage. On this unremarkable slope, north of the Blayney Meadows junction and south of Senger Creek, is the halfway point (170.5 km, 106 miles) for the John Muir Trail. Halfway up the slope, look to the south across the valley; the dark rocks down low in the valley belong to the Goddard pendant. The lighter colored rocks on the upper valley wall is the Mount Givens Granodiorite, which is probably one of the largest pluton in the Sierra Nevada batholith. The trail then ascends over Tioga till, passing just south of two moraine crests before fording Senger Creek, then continues to climb over Tioga till.

The trail climbs several perfectly developed recessional moraines, which forms a dam blocking in the Sally Keyes Lakes. On the isthmus between the lakes, composed of Tioga till (Qg), are numerous good campsites. To the east of Sally Keyes Lakes, several thick, light colored, pegmatite sills are remarkably continuous as they cut the slightly darker Bear Dome Quartz Monzonite (KJbd). Near the north end of Sally Keyes Lakes, after fording the inlet, are two W-NW striking, mafic dikes (1.5 and 2.5 m thick). The mafic dikes contain distinct 3-mm long hornblende crystals. The trail ascends to Heart Lake and around the east side where good exposures of the Bear Dome Quartz Monzonite (KJbd) are worth examining.

Bear Dome Quartz Monzonite (KJbd and KJbdf)

The Bear Dome Quartz Monzonite, Maps 23-24, is composed of biotite, minor hornblende, quartz, plagioclase, and potassium feldspar. The rock is heterogeneous with a northwest anastomosing (braided or lens-shaped pattern) foliation, and it has abundant small mafic inclusions and aplite dikes (Fig. 81). Internal fabrics of the pluton are complex, and deformation of the unit has never been studied in detail. The age is inferred to be older than the nearby Cretaceous plutons on the basis of the deformation, but the pluton has not been radiometrically dated. Bateman (1992) suggested that the intrusive is related to the volcanic rocks of the Goddard pendant, using the observations of Lockwood and Lydon (1975) that the wall-rock contacts are gradational. Bateman (1992) suggested a Jurassic age on the inference of the mafic dikes in the pluton belong to the Independence Dike Swarm, but as discussed earlier in the guide, the age of these dikes are difficult to interpret.

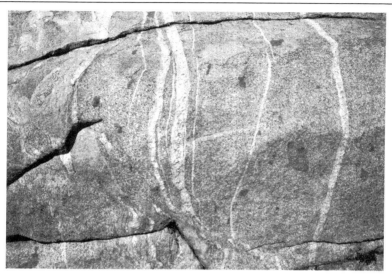

Figure 81. Outcrop photo of the Bear Dome Quartz Monzonite from along the John Muir Trail south of Selden Pass. Dark blebs are diorite inclusions (d) in a foliated granodiorite matrix. Vertical thin aplite dikes (A) were offset by a minor ductile fault (F) that diagonals upward from lower center to the right side of the photo. The aplite dikes are deflected as they approach the fault, giving a reverse relative sense of displacement. Vertical field of view approximately 2.5 m.

The trail crosses an internal contact inside the Bear Dome Quartz Monzonite with its outer fine-grained margin, and along this contact are inclusions of meta-volcanic rocks (Kvp). On the final climb up through the narrow valley to Selden Pass, most exposures show the heterogeneous nature of the margin of the Bear Dome Quartz Monzonite. The rocks record complex patterns within them, appearing mixed, swirled, layered, lens-shaped, and folded. These patterns all indicate the rock was deformed, mostly in a ductile fashion (the rock was hot enough to flow), and probably deformed while a good portion of rock was molten. The average composition granite melts at temperatures above 700° C, depending on the amount water present. Most of the plutons of the Sierra Nevada batholith probably contained as much 5% water in the magma. In general, the higher percentage of water in the rock the lower the melting temperature for the rock. Some of the deformation in the Bear Dome Quartz Monzonite may span the crystallization and cooling of the pluton and continued in the solid state.

Selden Pass (3,322 m, 10,900', Map 24, waypoint 334063E, 4128475N)

Selden Pass is a narrow notch in a subdued ridge and is composed of the Turret Peak Quartz Monzonite (Ktp). The east-facing wall just west of the pass has some irregular-shaped meta-volcanic rock inclusions (they are darker colored than the surrounding Ktp). The view to the north of Selden Pass overlooks the deep blue Marie Lake and the Bear Creek drainage (Fig. 82). Marie Lake was not filled in by post-glacial sediment during the last 10,000 years, and because the basin it occupies is relatively small and the amount of sediment delivered to the lake is also minor, the lake will probably keep its present form for the next 20,000 years. The prominent dark colored peak on the skyline to the north is Red Slate Mountain, composed of hornfels (a fine-grained metamorphosed sediment) belonging to the Mount Morrison pendant. To the south of Selden Pass is Heart Lake, and beyond this the drainage containing the South Fork of the San Joaquin River.

From Selden Pass, the trail descends northward and circumnavigates Marie Lake along the west side. Near the outlet to Marie Lake (Map 24), the trail overlies the Turret Peak Quartz Monzonite (Ktp). Both the west and east shores of Marie Lake (the southern 2/3rd's) are in the Turret Peak Quartz Monzonite.

Figure 82. View north of Selden Pass towards the glacial carved basin of Marie Lake. The low ridge to the left of the lake contains small screens of metamorphosed volcanic rock.

> **Turret Peak Quartz Monzonite (Ktp)**
> The Turret Peak Quartz Monzonite, Maps 23-24, is a porphyritic textured intrusion similar to Mono Creek Granite. The Turret Peak Quartz Monzonite is lighter colored than the Lamarck Granodiorite, has less biotite, and contains abundant 1- to 1.5-cm long k-spar phenocrysts. The islets in the lake are composed of Ktp. On the southeast side of Marie Lake is a large stock of Lamarck Granodiorite (Kl) intruded into the Ktp.

The trail leaves the Turret Peak Quartz Monzonite near the outlet to Marie Lake. Also near the outlet to Marie Lake, the trail crosses a septum of metamorphosed rhyolite tuff (JTrmv), which contains interbedded lenses of dark hornblende hornfels, and rare meta-sandstone and meta-conglomerate beds. These rocks were produced in a continental volcanic arc that formed a belt of volcanic centers parallel to the Jurassic coastline. The rocks have a strange mottled pattern to them, having discontinuous black and white layers. On the other side of the septa the trail is on Lamarck Granodiorite (Kl). The foliation in the Lamarck Granodiorite is at a low angle in these outcrops. After going around the east side of the garden-like Rosemarie Meadow, the trail fords the West Fork of Bear Creek and crosses a bit of Tioga till (Qg). The trail descends to the contact between the Lamarck Granodiorite and the Lake Edison Granodiorite (Kle).

The contact between the Lamarck and the Lake Edison Granodiorites is marked by a prominent, steeply dipping, foliation and highly stretched mafic inclusions on the trail. It is difficult to find a place where one can place a hand on the contact and say "here it is." If you wish to do this, try looking on the slabs to the east of the trail. When the trail enters the fine-grained margin of the Lake Edison Granodiorite watch for a decrease in grain size, a slight lightening of the rock, and lesser percentage of hornblende.

In this region, numerous EW- to NE-striking minor faults offset the contact between the Lamarck and Lake Edison Granodiorites, and also crosscut felsic dikes with left-lateral offset (Lockwood and Moore, 1979; Martel and others, 1988; Burgmann and Pollard, 1994). Most of these faults display en echelon patterns, and complex deformation at the overlap regions between adjacent faults. Lockwood and Moore (1979) suggested that the regional sets of systematic NE-striking left-lateral faults and more NS-striking right-lateral faults and joints represent tensional stresses related to Cenozoic development of the Basin and Range province. However, hydrothermal minerals deposited along the faults give an age of about 75-79 Ma (Segall and others, 1990), showing that most joint sets probably pre-date the Cenozoic. Bergbauer and Martel (1999) evaluated joints in this region to be caused by contraction related to pluton cooling. Deformation here is similar to EW-striking faults in the Mount Whitney Intrusive Suite (Lockwood and Moore, 1979). The similarities of joint and fault orientations across numerous plutons, and the age of formation cited above, strongly suggest this joint and fault set is a tectonic feature (Tikoff and Saint Blanquat, 1997) and is not related to pluton cooling or to uplift of the batholith. Many of the curvi-planar joints are NE-striking and have minor apparent left-lateral offset (Lockwood and Moore, 1979). Along Bear Creek, Burgmann and Pollard (1994) also described the pattern of fracture linkages, noting the connection by splays or wing cracks and concentration of hydrothermal fluids at these intersections. Similar fractures and faults are associated with the Rosy-Finch shear zone (Pachell and Evans, 2002; Pachell and others, 2003).

Lake Edison Granodiorite (Kle)

The Lake Edison Granodiorite, Maps 24-26, ~90 Ma (Tobisch and others, 1995), is equigranular, medium-grained, overall gray, hornblende-biotite-granodiorite. The granodiorite contains abundant sphene. This pluton is elongated in a northwest direction and is part of the John Muir Intrusive Suite (Fig. 65). The Rosy-Finch shear zone, a ductile fault, deformed the pluton while it was still hot, and possibly caused the bottleneck geometry at the pluton's center (Tikoff and Teyssier, 1992). Lockwood and Lydon (1975) subdivided a finer-grained facies (Klef) to the pluton.

The trail descends to ford Bear Creek at Upper Bear Creek Meadows (waypoint 334626E, 4132918N). This stream crossing can be difficult under high flow conditions. Farther downstream, glacially carved slabs, just north of the trail junction for Italy Pass, expose numerous thick aplite dikes and diorite inclusions (Fig. 83). An excellent exposure of compositional layering or schlieren can also be found on these glacial polished slabs (Fig. 84). The alternating light and dark minerals suggest rhythmic crystallization and perhaps mechanical sorting of the mafic minerals by magma flow.

Figure 83. Well-rounded diorite inclusions in the Lake Edison Granodiorite. Lens cap for scale.

Figure 84. Schlieren laying in the Lake Edison Granodiorite, crop out along Bear Creek in glacial polished horizontal slabs near the John Muir Trail. Note convergence of mafic bands towards the contact with the lighter homogeneous granodiorite on the left side of the photo. Photograph covers about two meters of width.

The trail follows Bear Creek downstream for 5.6 km (3 1/2 miles) to the lower trail junction for **Lake Thomas Edison (ENTRY POINT, Map 25, waypoint 332909E, 4137190N)**. Once the trail begins the long, arduous, climb up south side of Bear Ridge, the vegetation mainly consists of aspen, cedar, sage, and mule ears. Several springs crowded by Columbine, lupine, and Queen Anne's Lace are also passed on the climb. The higher the trail climbs, the fewer outcrops are seen of the Lake Edison Granodiorite, for it is completely draped by till deposits. Near the top of the ridge two bouldery moraine crests composed of Tahoe till (Qta) are crossed over, then the trail comes to the upper junction for **Lake Thomas Edison (ENTRY POINT, Map 26, waypoint 330943E, 4138829N)**. After the junction, the trail crosses the contact between Qta and possible Sherwin till (Qs).

The trail gently ascends over the rounded top of Bear Ridge, an area originally mapped as being volcanic rocks. Much of the rounded volcanic boulders, occasionally mixed with large granitic boulders, may indeed be Sherwin till composed of volcanic material derived from flows at Volcanic Knob 1 kilometer (~1.6 miles) to the east along Bear Ridge. At a minimum, the granitic erratics on top of Bear Ridge probably pre-date the Tahoe Stage. I have assigned these rocks along the trail to the Sherwin till because of the lack of morainal crests, the rounded form of the deposit (similar to the Sherwin till at Bighorn Plateau at the south end of the JMT), and its high elevation. Birman (1964) identified granitic erratics on the upland southwest of Volcanic Knob as Sherwin stage deposits.

The rock type at Volcanic Knob is an olivine basalt that erupted at about 3.6 ± 0.1 Ma (Dalrymple, 1963). These volcanic rocks are the next youngest extrusive igneous rocks along the JMT to those of the Devils Postpile National Monument to the north. This location, and the Devils Postpile flows, are the only two places along the trail having young, extrusive igneous rocks. The top of Bear Ridge is forested by Lodgepole pines, these give way to Red Fir and then White Fir and Aspen on the descent down the north side of the ridge. The north side of Bear Ridge is covered by talus and Tioga till, over which the trail descends many switchbacks down to the next drainage. Two thirds of the way down Bear Ridge, the trail descends into an Aspen grove, and on the north side of the grove is a glacially carved ridge adjacent to the trail. The low ridge is composed of the fine-grained portion of the Lake Edison Granodiorite.

At the base of Bear Ridge the trail crosses an unnamed stream, and along the south side of the trail, just pass the ford, several small cold springs on the margin to Quail Meadows have built up grass covered mounds of calcium carbonate. The rock around the springs is orange colored because of growth of algae. Several other springs are in the area near the trail. I observed one spring coming out of a vertical joint half way down a small, rock cliff face to the northwest of the trail. Near the Mono Creek crossing, just to the south, several excellent campsites on a low glacially carved ridge, paralleling the creek, are flat and have good trees for counter balancing food in.

The trail crosses over Mono Creek by a solid bridge and comes to a trail junction for **Lake Thomas Edison (ENTRY POINT, Map 26, waypoint 329743E, 4142191N)**. Mono Creek drains into Lake Thomas Edison, a man made reservoir finished in 1954 by the Southern California Edison Company. Tioga stage glaciers extended as far as Lake Thomas Edison, including terminal moraines that are now submerged beneath the lake (Birman, 1964).

The ascent up Mono Canyon passes over several small igneous rock bodies (Kqmp and Kjs). These units are difficult to distinguish from one another, especially because the trail is mostly on cover. Near the confluence of north fork and Mono Creek the trail is largely in talus, and Qal. The trail fords to the south side of Mono Creek, which can be difficult during high water, and then makes a steep ascent over rocks belonging to Graveyard Peak Leucogranite (Kjg).

> **Graveyard Peak Leucogranite (Kjg)**
>
> The Graveyard Peak Leucogranite, Map 26-27, 99 Ma (Stern and others, 1981), is light colored, medium-grained, equigranular, and white when fresh (orange on weathered surfaces). **Leuco** means light colored. Hornblende is common in this unit. The rock contains abundant mafic inclusions, is locally foliated, and it is cut by numerous aplite dikes.

The trail climbs to the **Mono Pass trail junction (ENTRY POINT, Map 27, waypoint 331031E, 4148303N)**. Near the 8,680' elevation, at glacially carved slabs, the trail intersects an intrusion here named the **Pocket Meadow Quartz Monzonite (Kpm)**. The rock is lighter colored than Kjg. This rock unit relatively thin, and the trail completely crosses it in 150 m (~450 feet). The Cretaceous age was inferred by mapping of Lockwood and Lydon (1975), but radiometric dating is unavailable. The trail climbs to intersect the contact of the Mono Creek Granite (waypoint 331084E, 4144164N), which has more quartz and abundant k-spar megacrysts, and is coarser grained. The contact indicated on the map is difficult to locate precisely on the ground because the exposure is poor and lichen covered.

> **Mono Creek Granite (Kmo)**
>
> The Mono Creek Granite, Maps 27-33, 88 Ma (Bateman, 1992), crystallized with a similar rock texture and composition as the Cathedral Peak and Whitney Granodiorites. It is the youngest member of the John Muir Intrusive Suite. The average composition is 26% quartz, 27% potassium feldspar, 41% plagioclase, and 6% accessories (mainly biotite, hornblende, and sphene). It has a porphyritic texture defined by large euhedral potassium feldspar crystals hosted in a coarse-grained matrix.
>
> The Mono Creek Granite is a good example of a pluton that had more than one process operating during its formation. On the east side of the pluton is a bulbous form that extends to the east into the Round Valley Peak Granodiorite (Fig. 65). This lobate shape was cited as evidence for forceful emplacement of the pluton, or expansion of the pluton, pushing aside the Round Valley Peak Granodiorite (Bateman, 1965b). Either concurrent, or subsequent to this, the Rosy-Finch shear zone was active, possibly providing room for the pluton to intrude towards the northwest, and definitely stretching and deforming the pluton (Tikoff and Teyssier, 1992). This combination of forceful intrusion and right-lateral shearing may have been working throughout the rest of the John Muir Intrusive Suite, resulting in a series of similarly oriented elongate plutons. The shape of the deformed plutons may be linked directly to the motion of the oceanic crust that was being shoved beneath the western edge of North America, the same crust that was melting to eventually form the granitic rocks of the Sierra Nevada.

The trail climbs to Pocket Meadow and then fords to the north side of Mono Creek (again sometimes a difficult crossing during high water). Next, after steep climb up the north valley wall the trail traverses near the base of a waterfall, and then continues to climb many short steep switchbacks. Looking to the east, up the canyon, the pronounced U-shaped valley once held the Mono Creek glacier. In many ways, this canyon is similar to Yosemite, yet on a smaller scale.

The trail skirts around the south side of a small meadow and then ascends along Silver Pass Creek, eventually fording to the north side. The valley containing Silver Pass Creek is U-shaped, and littered by several large Tioga stage glacial erratics (Fig. 85). The glacier that once ground away the floor and sides of this canyon flowed to join the larger ice flows in Mono Canyon. As the trail climbs up to Silver Pass Lake, to the southwest is an arete composed of the Graveyard Peak Leucogranite (Kg). It is cut by numerous sub-horizontal aplite sills from the Mono Creek Granite. The trail circles around the east side of Silver Pass Lake, passing over several patches of Tioga till, abandoned debris of the last glacier, but someday it will probably once again be moved down hill by ice. A short climb over Tioga till brings one to Silver Pass.

Figure 85. Photograph of glacial erratics in the Silver Creek Valley to the south of Silver Pass.

Silver Pass (3,334.3 m, 10,940', Map 27, waypoint 330015E, 4148303N)

Looking south from Silver Pass, the drainage of Silver Pass Creek heads south to join the westward flowing Mono Creek. The ridge south of the Mono Creek Valley is called Bear Ridge. Darker-colored barren rocks are visible along the top of the ridge, forming bald patches composed of young volcanic flows. The skyline of Bear Ridge defines the gently sloping profile characteristic of the west side interfluves along the Sierra Nevada. Beyond Bear Ridge, the prominent low saddle is Selden Pass.

The view to the north from Silver Pass overlooks Cascade Valley. Beyond the valley are the darker meta-volcanic rocks of the Mount Morrison pendant. The two prominent peaks to the distant north are Ritter and Banner, both are part of the Ritter Range pendant (Fig. 86). The Mono Creek Granite extends from Silver Pass to the south end of the Ritter Range pendant. To the right of Banner peak is a flat-topped lower summit called Donohue Peak. The JMT climbs over Donohue Pass to the west of this latter peak.

Figure 86. View north from Silver Pass towards the Ritter Range pendant on the horizon. Prominent peaks in the background are Ritter and just to the right is Banner, both are part of the metamorphosed volcanic rocks in the Ritter Range pendant. On the right side of the photo, the dark rocks are the metamorphosed volcanic rocks in the Mount Morrison pendant.

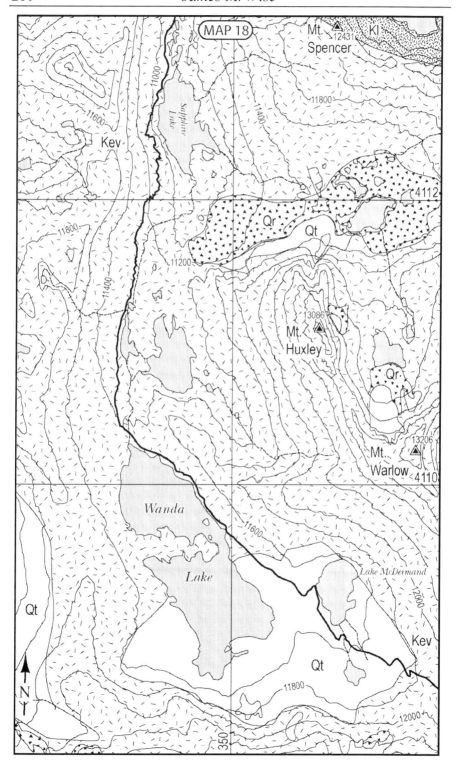

Geology of the John Muir trail

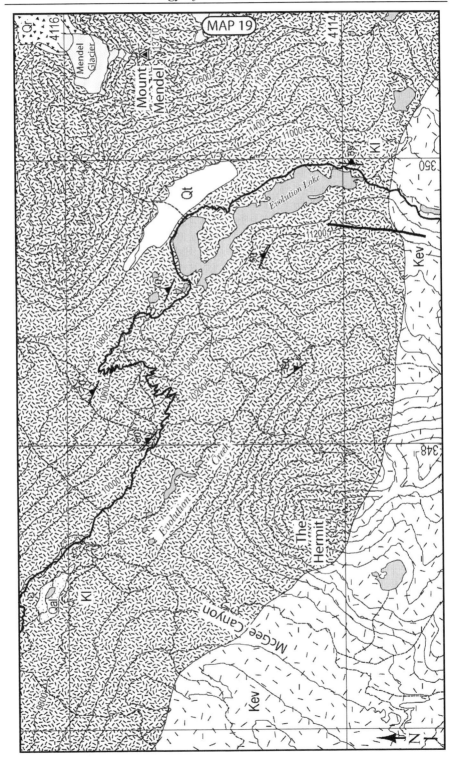

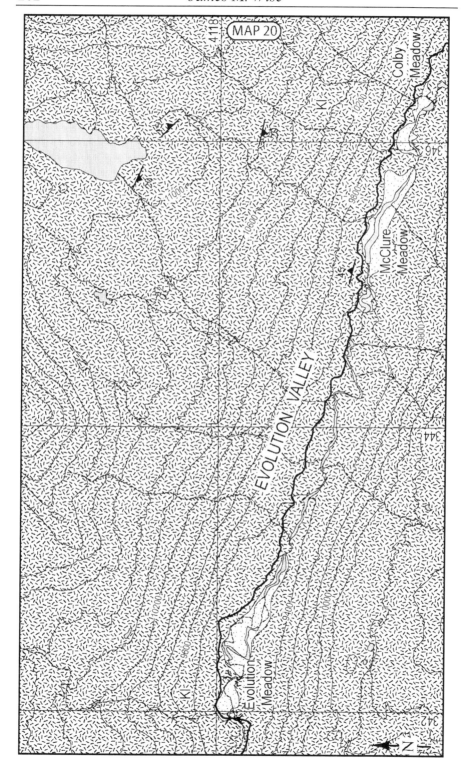

Geology of the John Muir trail 203

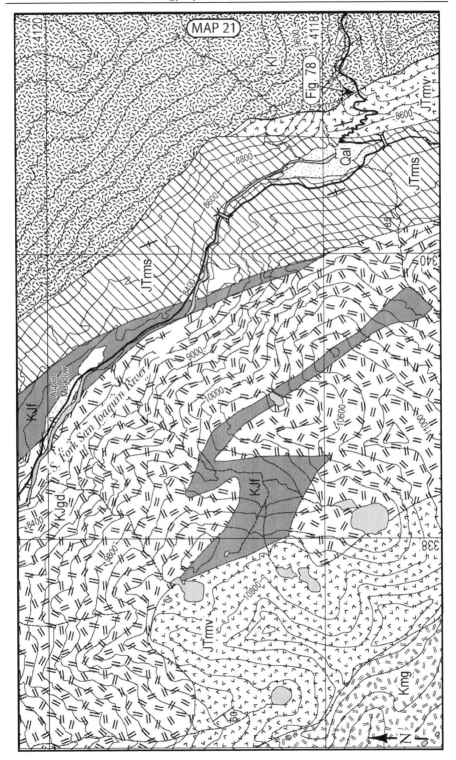

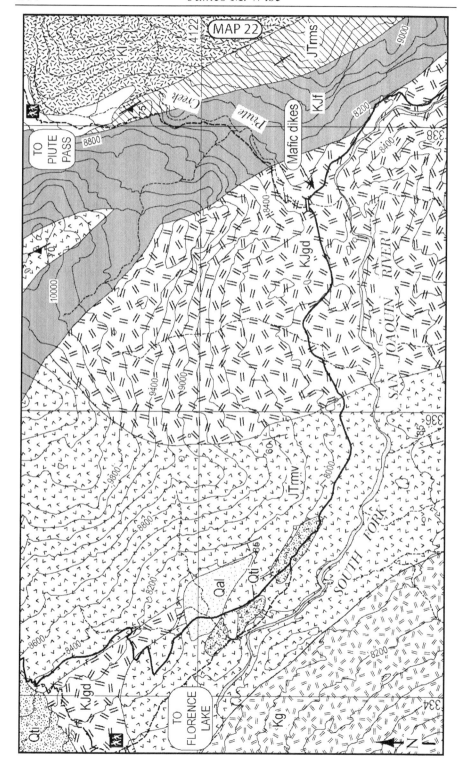

Geology of the John Muir trail

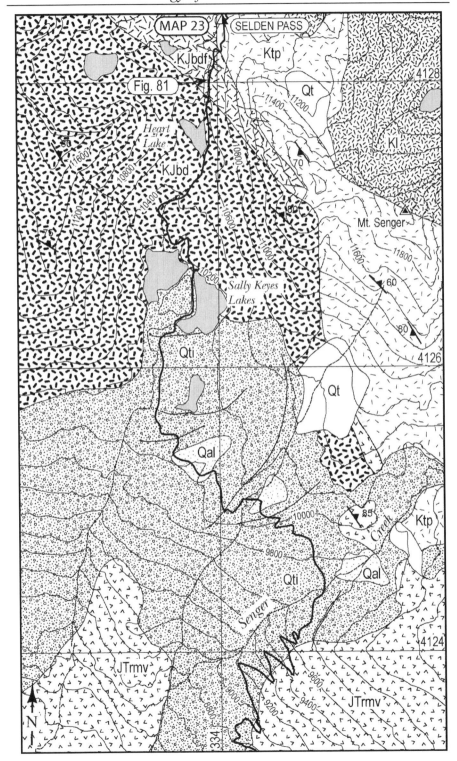

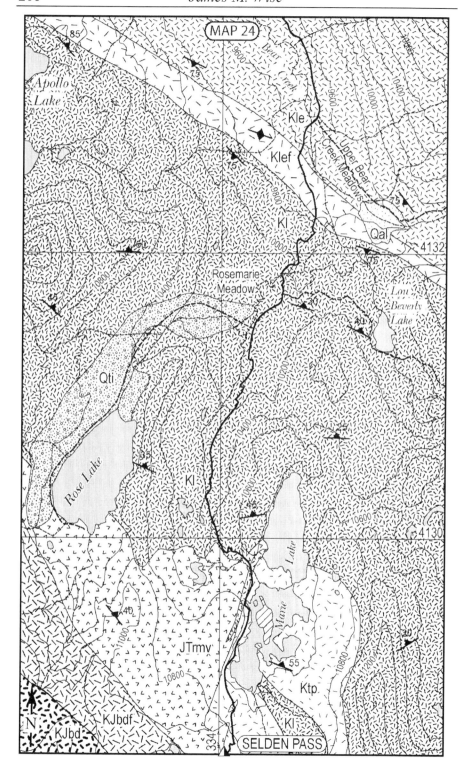

Geology of the John Muir trail

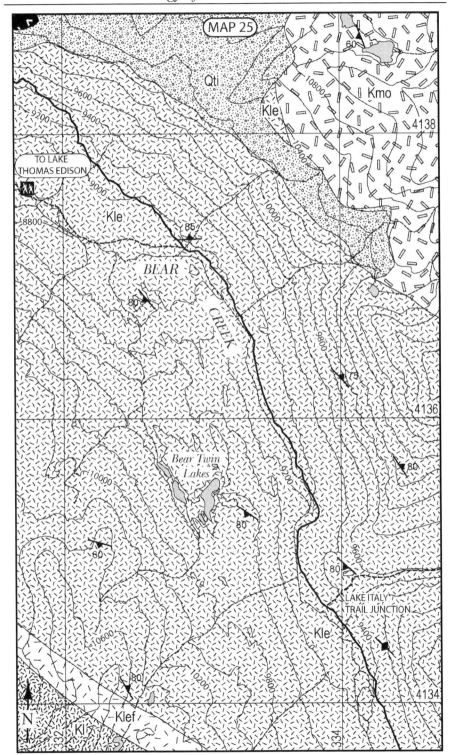

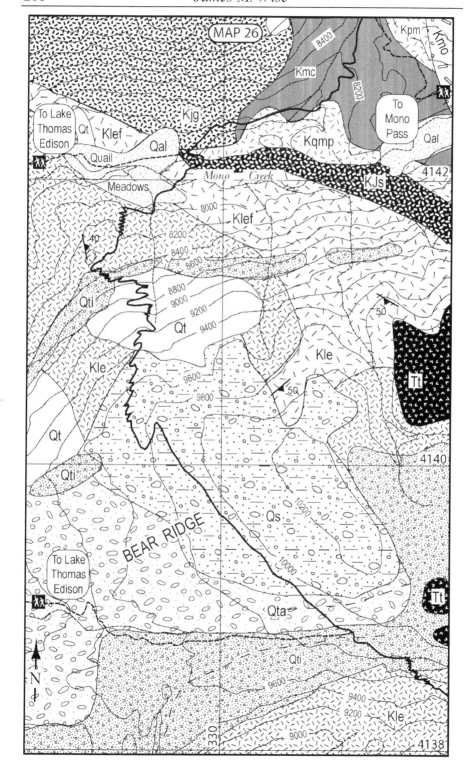

Geology of the John Muir trail

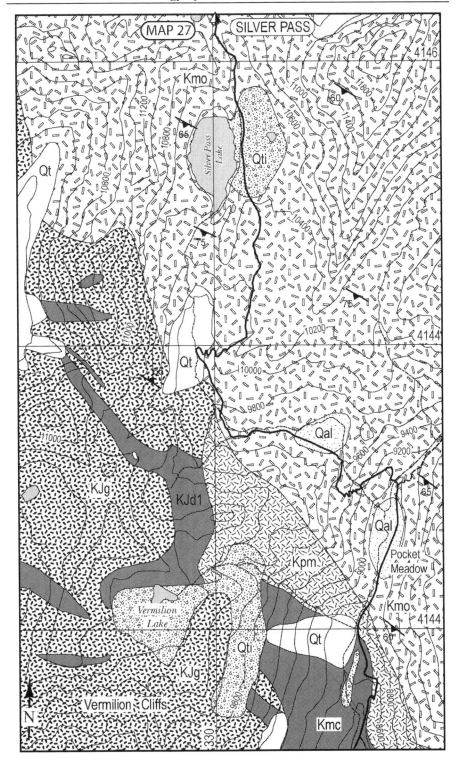

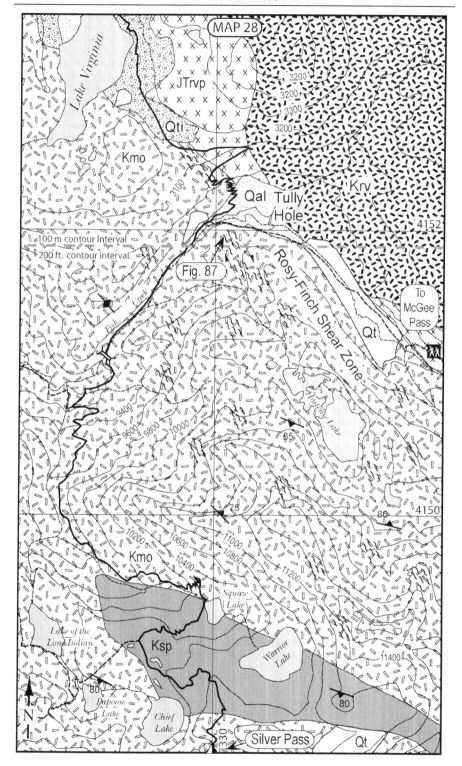

CHAPTER 8

SILVER PASS TO DEVILS POSTPILE

Access: Devils Postpile, Duck Lake Pass, McGee Pass, Mono Pass, and Fish Creek trail.
Distance: 37.8 km (23.5 miles).
Maps: 28-32.

The Silver Pass to Devils Postpile segment covers the Mono Creek Granite, metamorphosed volcanic rocks of the Mount Morrison pendant, the Rosy-Finch shear zone, and the recent volcanic lava flows of Devils Postpile National Monument.

The John Muir Trail descends north from Silver Pass, crossing the contact between the Mono Creek Granite and the Silver Pass Granite (Ksp) at the initial steep part of the descent. The contact was deeply covered in snow at the time this guide was being researched. The trail descends a ridge overlooking Chief Lake, and then reaches a trail junction for Goodale Pass (waypoint 329467E, 4149124N).

> **Silver Pass Granite (Ksp)**
>
> The Silver Pass Granite, Map 28, formed a small NW-SE trending elongate body of fine- to medium-grained biotite quartz monzonite and granite. The Silver Pass Granite is different from the Mono Creek Granite because it does not have the megacrysts of k-spar, it is finer-grained, and it contains more quartz and biotite. Overall, the rock has a clean and crystalline appearance to it, because it is exposed in glacier-polished slabs and since it contains abundant quartz grains. The unit has not been dated.

From the Goodale Pass trail junction, the JMT descends eastward to Squaw Lake. Near the outlet to Squaw Lake, well developed glacier striations and polish are present. At the outlet to Squaw Lake, the trail crosses the intrusive contact between the Silver Pass Granite and the Mono Creek Granite. The contact is more or less directly under the lake outlet. The rock containing large potassium feldspar crystals is the Mono Creek Granite. The trail descends along the drainage from Squaw Lake to the trail junction for Fish Creek, which leads down Cascade Valley and eventually to Devils Postpile. The JMT climbs up the south side of Fish Creek a short distance to a bridge, crosses the creek, and then follows it upstream along the north side of the valley to Tully Hole (Map 28). Once the trail crosses the steel and wood bridge over Fish Creek, the Mono Creek Granite is intensely foliated (parallel fabric/aligned minerals).

At the trail junction for McGee Pass, one can take a short detour, walking southward across the log bridge to the glacially carved slabs south of Fish Creek to examine some of the best exposures of the Rosy-Finch shear zone (RFSZ) along the JMT. The RFSZ is a Late Cretaceous right-lateral shear zone that probably in part accommodated the pluton emplacement of the John Muir Intrusive Suite (Tikoff and Teyssier, 1992). Some of the outcrops have folded aplite dikes that contain a foliation, and the k-spar megacrysts in the granite have a foliation wrapping around them that is defined by biotite grains (Fig. 87). This texture is called **mylonitic**, which is a result of ductile deformation. For those not willing to try this slight detour, the talus blocks just west of the trail junction illustrates the same mylonitic texture. The trail for McGee Pass has about 20.9 km (13 miles) to reach the McGee Creek trailhead.

The trail climbing up north valley wall above Tully Hole has many switchbacks. Looking back down at Tully Hole over the climb, the meadow is lined by a meandering stream, a channel pattern typical of larger river systems of continental lowlands. The uppermost switchbacks above Tully Hole traverses over dark gray metamorphosed volcanic rocks (JTrvp) belonging to the westernmost side of the Mount Morrison pendant. Near the top of the switchbacks, the trail crosses scattered outcrops of yellow-stained Mono Creek Granite. These outcrops are marked by a intense vertical foliation, the long axes of k-spar megacrysts are aligned, and sills of aplite record a well developed foliation, indicating the rocks are part of the Rosy-Finch shear zone.

Figure 87. Mylonite texture in deformed Mono Creek Granite in the Rosy-Finch shear zone at Tully Hole. Large white areas are potassium feldspar crystals. Note the foliation defined by segregated biotite. Lens cap for scale.

Mount Morrison pendant

The Mount Morrison pendant has some of the oldest sedimentary rocks preserved in the Sierra Nevada. The western 1/3 of the pendant is made up of Mesozoic metamorphosed volcanic rocks and the eastern 2/3 contains Paleozoic marine deposited meta-sedimentary rock (Fig 88, Maps 28-30). The pendant was intruded by both Triassic and Cretaceous intermediate composition plutons. The north side of the pendant is cut by the Pleistocene Long Valley Caldera, and to the east it is bounded by the active Hilton Creek range front normal fault. The pendant has a complicated history of Early Mississippian thrusting, Early Triassic folding, and Late Triassic left-lateral strike-slip faulting (Rinehart and Ross, 1964; Wise, 1996; Greene and others, 1997; Stevens and Greene, 1999). Subsequently, the pendant was contact metamorphosed and altered by igneous fluids in the Late Cretaceous. Even so, the original deformation fabrics preserve a record of the Early Mesozoic Sonoma orogeny superimposed folding on the mid-Paleozoic Antler orogeny thrust repeated stratigraphy, and later faults related to a Triassic continental truncation event. Both the Antler and

> Sonoma orogenic belts diagonal across the state of Nevada and probably once formed significant mountains. The rocks of the Mount Morrison pendant, and the adjacent pendants at the northern Ritter Range and Saddlebag Lake, represent the farthest west preserved record of these Paleozoic mountain belts. The metamorphosed volcanic rocks along the west side of the pendant are in intrusive contact with the Mono Creek Granite and was deformed in the right-lateral Rosy-Finch shear zone.

The trail passes through a saddle and passes over Qal and Tioga till as it winds around the east side of Virginia Lake to the inlet. At the inlet of Virginia Lake, the rounded boulders are composed of the Round Valley Peak Granodiorite (Krv) transported from the east by the glacier that once flowed through the valley. Virginia Lake sits half and half on top of the contact between the Mono Creek Granite (Kmo) and meta-volcanic rocks of the Volcanic Series of Purple Lake (JTrvp) (Fig. 89). The trail climbs north from the lake over Tioga till to a saddle. As the trail passes through the saddle, just to the south are blocky rock glacier deposits and to the north steep outcrops of JTrvp. The trail descends to Purple Lake on meta-volcanic rocks and then over Tioga till.

Geology of the John Muir trail

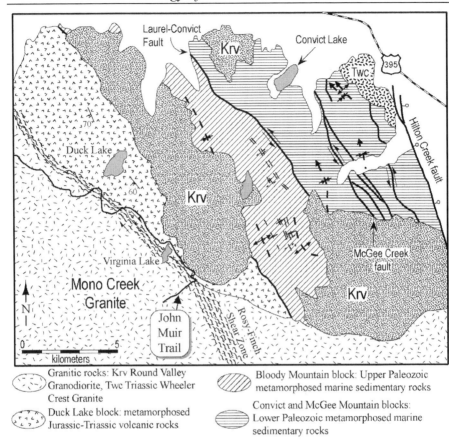

Figure 88. Simplified geologic map of the Mount Morrison roof pendant (from Wise, 1996).

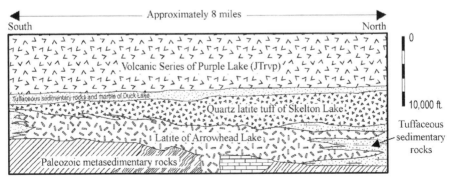

Figure 89. Generalized cross section of the volcanic rocks on the west side of the Mount Morrison pendant (from Rinehart and Ross, 1964). The John Muir Trail only passes over the upper part of the Volcanic Series of Purple Lake.

Round Valley Peak Granodiorite (Krv)

The Round Valley Peak Granodiorite, Maps 28-29, 89 Ma (Stern and others, 1981), is part of the John Muir Intrusive Suite (Fig. 65). The rock is medium-grained, equigranular, containing euhedral plagioclase, biotite, hornblende, and envelops small diorite inclusions. This rock lacks the megacrysts of potassium feldspar typical of the Mono Creek Granite.

Volcanic Series of Purple Lake (JTrvp)

The Volcanic Series of Purple Lake, Maps 28-30, are metamorphosed Jurassic-Triassic volcanic rocks made of pyroclastic material (a violent fragmental extrusive rock). The JTrvp is the youngest unit in a related stack of layered volcanic flows, and it unconformably overlies (was deposited on an erosion surface) the tuffaceous meta-sediments of Duck Lake (Fig. 89). The volcanic units are lenticular-shaped on a scale of hundreds of feet (Rinehart and Ross, 1964). The rock is dark to light gray, containing K-spar 1-2 mm in size, porphyritic in texture, and has discontinuous interbeds of marble and siliceous hornfels. These rocks erupted in a volcanic complex at the same time as the Jurassic volcanic centers seen along the JMT to the north at the Ritter Range pendant, and to the south in the Goddard pendant. The volcanic rocks at all three locations during the Late Jurassic were probably continuous, forming a volcanic arc, or mountain range, similar to the Andes. Irregular green pods of epidote are common and represent a chemical replacement of the rock during metamorphism. Also, the rock is marked by an intense near vertical foliation and slaty cleavage. Bergeron (1992) studied the fabrics in the meta-volcanic rocks, determining they have undergone two deformations. The second deformation can be matched to the Late Cretaceous Rosy-Finch shear zone. The earlier deformation, most likely responsible for tilting of the beds to near vertical, may be from Late Jurassic terrane collisions at the subduction zone to the west.

Purple Lake is blocked by recessional moraines from the retreat of a Tioga stage glacier. To the east of Purple Lake, up the valley, the light-colored high cliffs are composed of the Round Valley Peak Granodiorite (Krv). The trail recrosses the intrusive contact between meta-volcanic rocks (JTrvp) and the Mono Creek Granite (Kmo) near the west side of the valley floor (waypoint 327489E, 4155033N). Several septa of JTrvp are contained in Kmo near the contact. The trail climbs uphill along a ridge then descends and contours towards the valley containing the Duck Lake drainage.

The trail descends through a chute, and recrosses into the Volcanic Series of Purple Lake (JTrvp). The contact is dramatically exposed just east of the trail where numerous felsic dikes intruded into the metamorphosed rocks. Some of the dikes are folded and others are boudinaged (stretched into pieces) as seen in a vertical direction of exposure (Fig. 90). The trail descends to the **Duck Lake Pass trail junction (ENTRY POINT, Map 29, waypoint 326099E, 4156208N),** which leads to Mammoth Lakes. The JMT then drops down to ford the Duck Lake drainage.

The contact between the Mono Creek Granite and the Volcanic Series of Purple Lake is intersected on the west side of the ford. The intrusive contact is beautifully exposed in the glacially carved valley floor. The Mono Creek Granite sent numerous dikes and sills of aplite and pegmatite into the metamorphosed volcanic rocks. Subsequently, these dikes and sills were folded (Fig. 91) possibly while the Rosy-Finch shear zone was active in the Late Cretaceous.

The trail climbs up from the valley, going over the Mono Creek Granite, and then over slopes covered by till composed of the Mono Creek Granite. The trail crosses the gradational contact between till composed of granite boulders and till primarily composed volcanic rocks near the 3,047 m (10,000)' elevation. Most of the boulders are rounded in shape, suggesting glacial transport of the material. The trail makes a gentle descent along the north rim of the valley. The Mono Creek Granite (Kmo) crops out just below this moraine deposit all along the steep cliffs to the south of the trail.

Figure 90. Vertical outcrop of boudinaged (pulled-apart) aplite dikes that are hosted in deformed metamorphosed volcanic rock (dark areas in the photograph) along the western margin of the Mount Morrison pendant. In this view, the dikes are parallel to foliation in the metamorphic rock. The deformation here from the Rosy-Finch shear zone also created folding of similar dikes to the north, as shown in Fig. 91. Field of view is approximately two meters across.

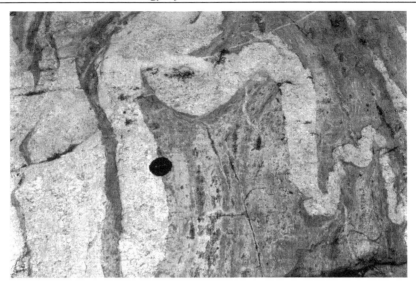

Figure 91. Photo of folded aplitic dikes of the Mono Creek Granite intruded into metamorphosed volcanic rock of the Mount Morrison pendant. The fold hinge surfaces of the dikes are parallel to the foliation. The dikes do not show internal foliation, indicating that they were folded while still partially molten. Outcrop view is in a glacially polished horizontal surface. Similar dikes are stretched and pulled apart vertically (Fig. 90). Lens cap for scale.

The trail meanders northward over hummocky terrain made up of Tioga till and descends to Deer Creek (Map 30). The trail turns left at an unmarked trail junction; the path heading to the east is not as well traveled and it dead-ends up the valley. The trail heads westward across flat ground, climbs a short hill, and then turns a low-profile point. When the trail reaches a small spring, which is 1.5 m (5') west of the trail, one may glance through the pines to the west to see a short reddish-colored cliff at the top of a ridge. This is the best exposure of Andesite of Pumice Butte (Qap) as seen from the John Muir Trail. This lava unit is part of the volcanic suite erupted in the Devils Postpile National Monument.

To the east of the trail, as one ascends a slight valley, are low-angle slabs of the Mono Creek Granite (Kmo). In the slabs, the k-spars phenocrysts are up to 2 cm in length, the rock contains coarse, irregular, grains of quartz, and very small, evenly distributed, biotite grains. The trail makes a gentle ascent along the covered contact (follows the stream bed) between Kmo and the Pliocene **Andesite of**

Pumice Butte (Qap- composed of cinder cones and flows; the lava is scoriaceous and non-porphyritic). The Andesite of Pumice Butte (Maps 30-31) predates all the known glaciations of the Sierra Nevada. Most of the Andesite of Pumice Butte is located at the hill to the west of the trail. The pumice and forest covered slopes to the west contain float (loose rocks in slope cover) of vesicular gray and red andesitic basalt belonging to Qap. Some of these fragments appear to be ejecta. To the east, high on a ridge, is an arete composed of Kmo named the Thumb. The trail crosses a pumice-covered flat with scattered small pines, over colluvium (Qal), and then passes through a slight saddle. The pumice most likely originated from eruptions of Mammoth Mountain to the north, which is an active volcano formed on the rim of the Long Valley Caldera.

The trail again executes a gradual descent towards the north in Tioga till (Qg), passing some low-angle slabs of Mono Creek Granite (Kmo) that are just to the east of the trail. Next, the trail bypasses Upper Crater Meadow (Map 31) to the south. Unfortunately, the trail was recently moved from its old route through the meadow, thereby trading verdant landscape for more dusty pumice slogging. After crossing a second sturdy wood bridge at the north side of Upper Crater Meadow is a trail junction for Mammoth Pass. The steep cliffs to the east are still in Kmo. The low hill to the north of the trail junction is composed of Tioga till (Qg).

The trail descends a gentle hill and crosses Crater Creek. From the creek, one may look to the north for a good overview of the north Red Cone (Fig. 92) and the south side of Mammoth Mountain. The two cinder cones are the second youngest rock formations crossed by the JMT (the youngest geologic events are the talus deposits formed by rock avalanches). The nearby stream banks are composed of Tioga till (Qg). The trail descends a few switchbacks to Crater Meadow and continues northwest across a stretch of flat ground. The trail fords Crater Creek near the base of the northern Red Cone.

Figure 92. View northward looking at the north Red Cone (Map 31). The cinder cone is composed of vesicular basalt and was erupted onto a glaciated granite surface. Mammoth Mountain is the background.

Basalt of the Red Cones (Qbr)

The Basalt of Red Cones, Map 31, is composed of cinder and flows of basalt. The rock is dark, has a fine-grained matrix, sprinkled with phenocrysts of plagioclase and olivine. The flows originated from the bases of the cones (volcanic vents), and were gas rich because the basalt is vesicular (contains gas bubbles). Near the ford to Crater Creek is a short side trail leading to the top of the northern Red Cone. Exposed along this trail are several volcanic features including an old lava tube **skylight,** flows made of **pahoehoe lava** (Fig. 93A), a **tree mold** (Fig. 93B), and the cinder vent. A skylight is a hole broken in the roof of a lava tube. The skylight exposed here has the lava tube plugged by basalt. A tree mold is an impression of a tree in a lava flow. The one on the south flank of the cone is in an upright position. Once the lava surrounds the tree, the tree is burned away leaving a hole in the lava. A vent is the source for volcanic eruptions of molten rock onto the Earth's surface. Both the cinder cones piled up more ejecta (projectiles of fragmented magma) on the northeast sides, indicating this was the dominant wind direction at the time of eruption. Most of the basalt flows originated from the south cone and cascaded down the valley wall. Both cinder cones are developed on top of the Mono Creek Granite (Kmo). These volcanic rocks were not glaciated, so they are younger than about 12,000 years. Also, they are partially covered by pumice that erupted about 500 years ago from Mammoth Mountain (Bailey, 1989).

Figure 93A. Photo of vesicular basalt in pillow lava-pahoehoe form in flow at the base of the north Red Cone. Lens cap at photo center for scale.

Figure 93B. Photograph looking down into a tree mold on the south flank of the northern Red Cone.

Summary of late Cenozoic Sierra Nevada Volcanism

Recent volcanic activity along the east side of the Sierra Nevada includes rhyolite domes near Mono Lake, basalt flows at Mammoth Lakes, the large Long Valley Caldera, basalt cinder cones and flows south of Big Pine and to the south of Lone Pine at Red Cone and Fossil Falls (part of the Coso volcanic field). All of the volcanic rocks are related to the active normal faults of Owens Valley. Fairbanks (1898) noted the close association between volcanism and faulting along the eastern escarpment early in the study of the Sierra Nevada. The faults provide the conduits or path for the magma to reach the surface. These volcanic rocks lie at the extreme southern end of the modern Cascade volcanic arc, which is defined by large andesitic stratovolcanos such as Mount Rainier, Adams, and Shasta. These latter volcanoes are products of subducting oceanic crust beneath western North America. However, once south of Mendocino along the coast of California, the plate boundary changes from subduction to the purely strike-slip motion of the San Andreas Fault. This means that all of the late Tertiary basaltic rocks along the eastern side of the Sierra Nevada are not products of

> subduction. Moore and Dodge (1980) suggested this tectonic setting explains the relatively enriched amount of potassium in the basalt flows, in what they called the San Joaquin-Kings volcanic field.

From the Red Cones, the trail descends drawn-out switchbacks into the South Fork of the San Joaquin River canyon. On the descent, some outcrops of the Mono Creek Granite are passed. The trail is mostly on lateral moraines from the Tioga stage. In fact, the longest switchback follows along one of the moraine crests (Map 31).

The trail passes below some short cliff bands composed of the Bishop Tuff (visible occasionally through the trees grown with bent trunks). Instead of scrambling up the hill to look at this rock, look for boulders along the trail that fell from the cliff. The boulders are reddish tan to gray, and marked by large lens-shaped pockets weathering into them. The pockets were formed by flattened pumice fragments that later chemically weathered out of the rock. While the tuff was still hot from its eruption, the overlying weight of the tuff smashed the pumice fragments; this texture is called **eutaxitic foliation**.

> ### Bishop Tuff (Qb)
> The Bishop Tuff, Maps 31-32, is about 760,000 years old pyroclastic rock erupted as a nearly continuous thick composite sheet of ash flow tuff deposits from here to the Long Valley Caldera in the east (Bailey, 1989). A refined age of 758,900 6 1,800 years before present was reported by Sarna-Wojcicki and others (2000) from averaging 70 analyses from material sampled at five locations. Since its formation, most of the tuff in the San Joaquin drainage was removed by glaciation. This deposit is important because it demonstrates that the bulk of the San Joaquin Canyon was carved out before the Tahoe stage glaciation. Remnants of the Bishop Tuff are also located east of Reds Meadow.

At the base of the switchbacks the trail passes nearby a green spring and moves into the region covered by the 1992 Rainbow forest fire. From the spring to the Rainbow Falls trail junction, the trail goes over the Andesite of Mammoth Pass for ~1.6 km (1 mile).

> **Andesite of Mammoth Pass (Qam)**
>
> The Andesite of Mammoth Pass, Maps 31-32, is a light to dark gray, aphanitic (very fine-grained), slightly vesicular, andesite lava flow hosting varying amounts of small plagioclase phenocrysts. The outcrops compose scattered rocky knobs protruding from the small hills on either side of the trail. Since the forest fire of 1992 most of these exposures were blackened, so the rock may have to broken open to examine it. The flows are massive and in places have poorly developed thick columnar jointing. The Andesite of Mammoth Pass is one of several volcanic units in a succession of flows in the Devils Postpile National Monument area that are listed below in chronological order. It is important to note that, from oldest to youngest, the volcanic compositions possibly evolved through time, becoming more mafic. This may hint at a process within the magma chamber deep beneath Devils Postpile National Monument.

At the junction for Rainbow Falls, one may take the Rainbow Falls optional trip. The trail to the northeast, junction located at waypoint 316746E, 4164523N, leads to the **Rainbow Falls trailhead and ENTRY POINT (waypoint 316804E, 4164844N)**. A stock trail leads to the Reds Meadow Resort where telephones, a general store, and the Mule House Café may be a valuable resource to the hiker. Farther to the north, at the east side of the Reds Meadow campground, is a free shower house with the hot water supplied from springs. Hot springs of the region is another indication of the recent volcanic activity. Shuttle buses provide transportation, approximately $8 for a round trip, to Mammoth Lakes where numerous services are available.

Sequence of events at the Devils Postpile National Monument (listed from youngest to oldest, ka = 1,000 years)

~500 years	Region blanketed by pumice erupted from Mammoth Mountain (Bailey, 1989).
<10 ka	Basalt of the Red Cones erupt.
~14-25 ka	Tioga stage glaciation.
<100 ka	Eruption of the Basalt of Devils Postpile.
~125-160 ka	Tahoe stage glaciation.
	Eruption of the Andesite of Mammoth Pass.
	Eruption of the Rhyodacite of Rainbow Falls.
~215-235? Ka	Mono Basin stage glaciers erode most of the Bishop tuff.
759 ka	Long Valley Caldera eruption and deposits the Bishop Tuff.
?	Eruption of the Basalt of the Buttresses.
~1 – 1.3 Ma	Sherwin stage glaciation.
>1.3-<2.6 Ma	McGee stage glaciation
3.1-3.5 Ma	Volcanics of Deadman Pass
~5 Ma	Uplift and erosion of the Sierra Nevada begins.

Two hundred feet northwest of the junction for the Rainbow Falls trail and trailhead, a low ridge exposure of the Rhyodacite of Rainbow Falls (Qrr) contains interesting textures. The rock is made of auto-brecciated flow material. Overall, the rock has a light purple fine-grained matrix and dark angular fragments containing white plagioclase phenocrysts. This exposure is probably part of the lava flow top that cooled and fractured into chunks while molten rock deeper in the flow was still moving.

Rhyodacite of Rainbow Falls (Qrr)

The Rhyodacite of Rainbow Falls, Map 32, is a more felsic lava flow (light-colored and silica rich) than any of the other volcanic rocks in the Devils Postpile Monument. It is a light gray on fresh surfaces and light tan on weathered surfaces. White colored phenocrysts of plagioclase are common. The rock matrix is aphanitic (very fine-grained). It has a prominent platy cleavage (breaks into thin sheets) that is commonly sub-horizontal. At the overlook for Rainbow Falls, the Andesite of Mammoth Pass is on the southeast side of the

> falls and overlies the Rhyodacite of Rainbow Falls on the north side. The Rhyodacite of Rainbow Falls is split by a well developed subhorizontal platy cleavage, whereas the Andesite of Mammoth Pass is massive in appearance.

Thirty five meters (115') down the JMT from the low ridge, a blocky outcrop of the Rhyodacite of Rainbow Falls (Qrr) lies two meters off the trail. Overall, outcrops of the Rhyodacite of Rainbow Falls are limited to the knobby exposures on top of the ridges. The JMT continues to the southeast, winding about gentle hills covered in pumice (an extremely vesicular textured light weight rock made up of volcanic glass). The pumice here is especially thick, causing the trail to be a dusty slog.

Just before the monument boundary, the trail meets the junction for the Devils Postpile-Rainbow Falls trail (waypoint 316258E, 4164817N); this is where the Devils Postpile optional route splits from the JMT. This optional route brings one up close to Devils Postpile, and a side trail leads to the top of the postpile. The Devils Postpile can be seen from the JMT by looking eastward across the valley and through the trees. From the junction, the JMT enters the Devils Postpile National Monument. A low, glacially scoured, ridgeline is passed over, and then a small patch of cover is crossed. At the east side of the San Joaquin River, the steep cliffs are composed of the Rhyodacite of Rainbow Falls. The platy cleavage at this outcrop dips about 30 degrees to the east. The trail crosses over a wood bridge then heads south for a short distance.

Devils Postpile optional route (Map 32)

The Devils Postpile optional route for the JMT stays on the east side of the river to pass directly beneath the Devils Postpile, and then it crosses the river and rejoins the JMT. At the base of the postpiles, the hexagonal columns of basalt are approachable. A short and steep trail climbs both the north and south sides of the hill, allowing one to examine the columnar joints in top view where the polygonal pattern is well exposed from Tioga stage glacial scouring. The columns and joint pattern formed by the solidification and cooling of the lava flow. You may note that not all the columns are vertical. In fact, several of them are curved. The direction of the columns relates to the pattern of heat loss during cooling of the lava flow, which if unequal can form curved columns. The trail heads north of Devils Postpile along the river and turns left at a trail junction to cross the San Joaquin River by a bridge. A short distance north of the junction is the north Devils Postpile trailhead, again the Mammoth shuttle may be caught here. In addition, near the trailhead is a small ranger kiosk where information is available on the Devils Postpile. Along the west side of the river, a small soda spring, orange-colored by algae, is 20 m (55 feet) from the trail. The trail climbs to the JMT/Devils Postpile trail junction (waypoint 315955E, 4166315N).

JMT trail description continued from south part of the monument

Once across the river, to the south of the trail on the opposite side of a small patch of cover. A low outcrop of tan colored rock, splitting into thin sheets, is made of the Rhyodacite of Rainbow Falls. The trail ascends over outcrops of Mono Creek Granite (Kmo) to enter a north-south trending corridor paralleling the joint pattern in the rock. In the corridor, abundant large potassium feldspar crystals are weathering out of the Mono Creek Granite. The trail passes a small hill composed of the Basalt of the Buttresses (Qbb) on the south side, follows a short switchback near the base of the slope, and then makes a gradual ascent along the thick pumice covered talus slope. The dark cliffs above and west of the trail are made of the Basalt of the Buttresses.

Basalt of the Buttresses (Qbb)

The Basalt of the Buttresses, Map 32, has roughly developed columnar jointing, is dark gray, and contains abundant olivine phenocrysts. This unit is not in contact with the other volcanic rocks in the Devils Postpile region, so its relative age is not known. It was speculated that it could be the oldest of the volcanic rocks around the monument (Huber and Eckhardt, 1985). This unit was deposited in the glacially

> carved floor, bracketing it to be less than 5 million years old, and it probably post dates the Sherwin stage glaciation.

Across the valley, the large white cliffs are composed of Mono Creek Granite (Kmo), and higher up are the barren reddish to tan slopes of Mammoth Mountain volcano. **Mammoth Mountain** (3,369 m, 11,053') is a cumulovolcano composed of quartz latite (a quartz poor and potassium rich type of rhyolite) flows and rhyolite domes formed between 50,000 and 200,000 years ago at the southern end of the north-south trending Mono-Inyo Craters volcanic chain (Bailey, 1989). Mammoth Mountain is also positioned along the western rim of the Long Valley Caldera (a feature not visible from the JMT). This volcano is very young, and seismically active caused by processes in the magma chamber below the mountain. The region is constantly being monitored by the United States Geological Survey in hopes of providing adequate warning to the populace of Mammoth Lakes in the event of a volcanic eruption. In the late 90's, an area of a high CO_2 and other volcanic gas emissions at the southeast base of Mammoth Mountain killed part of the pine forest. These are clear signs that molten rock is being redistributed in the subsurface.

The trail leaves the Basalt of the Buttresses near a shallow drainage and goes back over the Mono Creek Granite. The contact is covered. Its approximate position is obvious because the soil type changes color from dark gray to white. The contact surface between the lava flow and the underlying granitic rock is a nonconformity. A nonconformity is a preserved erosion surface, having either igneous or metamorphic rocks below the surface, and layered rocks on top of the surface (commonly sedimentary).

An excellent viewpoint for looking at the Devils Postpile is available about 15 m (~50 feet) off the trail at a small rounded granite point (Fig. 94). The trail continues northward on glaciated slabs of the Mono Creek Granite. Most slabs have rough-weathered surfaces, yet some retain small patches of glacial polish and striations. The trail descends to the junction with the Kings Creek/Devils Postpile trail (waypoint 315869E, 4165942N). When the trail reaches a large slab of granite, an excellent view eastward across the valley to the Pliocene Volcanics of Deadman's Pass, which form dark brown, horizontally layered volcanic flows, on the opposite valley wall, is worth looking for. These andesite flows, dated at about 3.0 to 3.5 Ma (data by Dalrymple, reported in Huber and Rinehart, 1967), preserve an older geomorphic surface beneath them that was called the Broad Valley stage by Matthes (1960). This surface gives evidence of paleotopography on the order of 2,000 feet of local relief in the Pliocene (Huber and Rinehart, 1967). The pyroclastic flows are part of the Long Valley Caldera system. Similar to the Basalt of the Buttresses, the Volcanics of Deadmans Pass once erupted into the San Joaquin drainage, and then were eroded away by glaciers. When the trail descends into a grove of Red Fir and Lodgepole pines it leaves the Mono Creek Granite (Kmo) for the Basalt of Devils Postpile (Qad). The contact between the two units is a nonconformity. The trail heads north through a forest, passing smaller, poorly exposed outcrops of the Basalt of Devils Postpile.

Figure 94. View westward from the John Muir Trail looking at the Devils Postpile.

Basalt of Devils Postpile (Qad)

The Basalt of Devils Postpile, Map 32, is a Late Pleistocene lava flow. It is light to dark gray, commonly porphyritic bearing phenocrysts of white, rectangular-shaped plagioclase and small olivine. Olivine is a glassy green mineral of commonly round or equant grain shape, and is one of the major minerals that compose the Earth's upper mantle. The Basalt of Devils Postpile forms well developed columnar joints. Columnar joints are regular spaced fractures developed from the cooling of a lava flow- when the lava cools it contracts, producing fractures in a polygonal pattern that form the columns in 3-D (Fig. 95). The best columnar joints are exposed at Devils Postpile proper, for which the volcanic flows are named after. The lava flow is younger than 760,000 years because it overlies the Bishop Tuff. The lava flowed out into the glacial carved San Joaquin Canyon, solidified, and then was mostly eroded away by subsequent glaciers, indicating that the lava is at least older than the Tioga stage. Dalrymple (1964) reported a K-Ar method age of about 0.9460.16 Ma from the mineral plagioclase. Most of the lava probably originated from near the base of Mammoth Mountain and flowed towards the west.

One hundred feet west of the next trail junction, the JMT crosses a small stream gully exposure of the Basalt of Devils Postpile. This outcrop has columnar jointing developed in it and the basalt contains over fifty percent of large, white, plagioclase laths (rectangular-shaped crystals). From the junction of the **Devils Postpile trail (ENTRY POINT, Map 32, waypoint 315955E, 4166315N)** and the John Muir Trail, the path heads to the left; signs clearly mark the JMT. The north end of the monument marks the end of this segment of the guidebook and the beginning of the Devils Postpile to Tuolumne segment.

Figure 95. Photograph of polygonal cooling joints in the Basalt of Devils Postpile. The outcrop is glacially polished and striated.

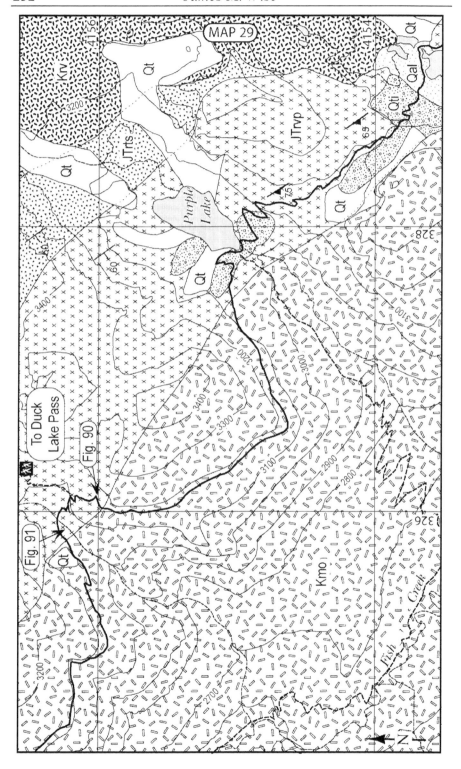

Geology of the John Muir trail 233

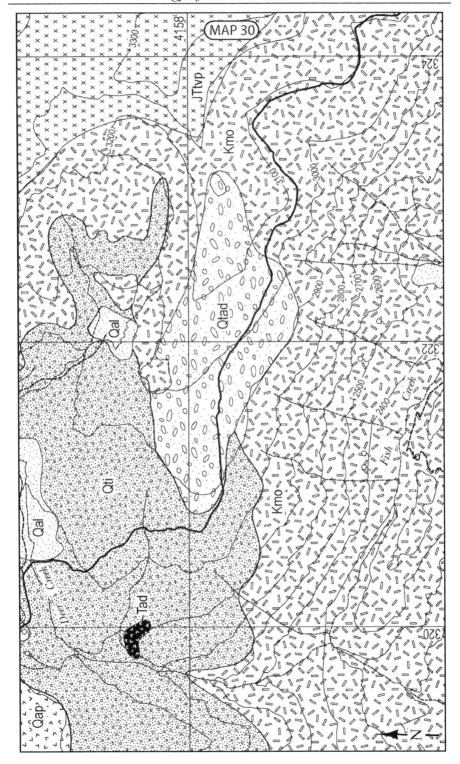

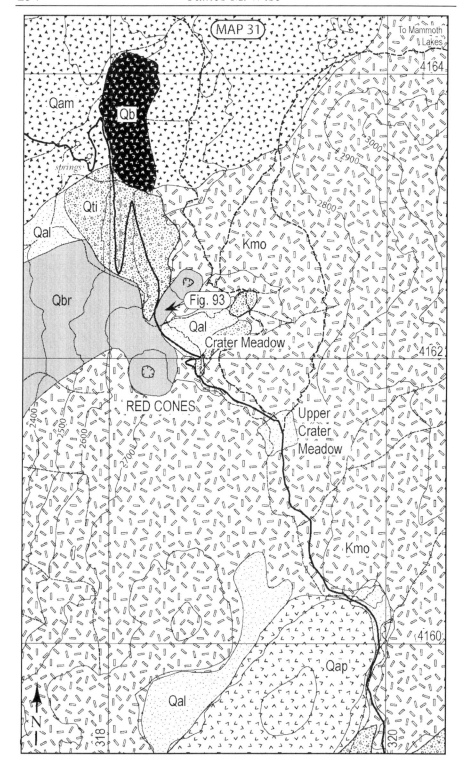

Geology of the John Muir trail

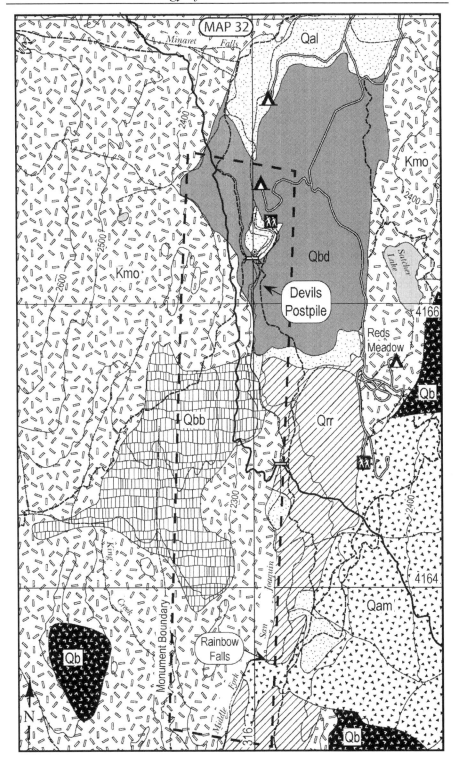

CHAPTER 9

DEVILS POSTPILE TO TUOLUMNE MEADOWS

Access: Trailheads at Devils Postpile, Agnew Meadows, Gem Lake, and Tuolumne Meadows.
Distance: 53.4 km (33.2 miles).
Maps: 32-39.

In this segment of the guide, the John Muir Trail leaves the John Muir Intrusive Suite and the young volcanic flows of Devils Postpile, crosses over the Ritter Range pendant, and then traverses almost half of the Tuolumne Intrusive Suite. The trail runs nearby to significant Cretaceous caldera deposits that account for the majority of the rocks in Mount Ritter and Banner. This guide segment begins at the sign marking the northern boundary to the Devils Postpile National Monument (either approached along the JMT or from the Devils Postpile trailhead). This section of the trail is in the Ansel Adams Wilderness area and Yosemite National Park.

Near the base of the west valley wall of the San Joaquin River, the trail crosses the nonconformity contact between the Basalt of Devils Postpile and the Mono Creek Granite (Kmo). The trail gradually climbs northward and converges with Minaret Creek, crosses it, then skirts the east side of a Johnston Meadow (Map 33). The trail leaves the meadow and ascends along the southwest side of a ridge then crosses a stream.

The trail crosses the intrusive contact between the Mono Creek Granite and metamorphosed volcanic rocks (JTrx) at a place west of the southern most Trinity Lake (waypoint 314425E, 4170531N). This is the northern end of the John Muir Intrusive Suite. The intrusive suite from end to end is about 105-km long. The contact between the Mono Creek Granite and the Ritter Range pendant is most likely steeply dipping. These metamorphosed volcanic rocks are part of the Ritter Range pendant (Fig. 96), which contains numerous deformation events. A good percentage of the pendant is Cretaceous age, composed of volcanic breccias erupted from the Cretaceous plutons. The volcanic rocks are made up of tuffs, tuff breccias, lavas, stocks, and some interbedded sedimentary rocks.

The trail shortly exits the meta-volcanic rocks (JTrx) and continues over a small body of diorite (KJd). The contacts are covered along the trail. The KJd crops out as low blocky hills several feet to the west of the trail. The rock is medium-grained and relatively fresh in appearance (not weathered). The diorite intruded the meta-volcanic rocks, indicating that it is younger than Triassic-Jurassic. Although, the rock has not been radiometrically dated, so the age assignment is an estimate. Near Trinity Lakes, the trail returns to the meta-volcanic rocks then passes Gladys Lake on the west side. The meta-volcanic rocks are massive, light to dark green and gray. They are best exposed at the saddle between Gladys and Rosalie Lakes (Map 33). The gray unit includes clear crystal fragments, indicating this is a metamorphosed volcanic tuff. The crystals are broken throughout the violent explosion by which tuffs are erupted from calderas. The trail circles around the east side of Rosalie Lake (in doing so crosses the same bed of JTrx twice), and then ascends slightly.

Ritter Range pendant

The metamorphosed volcanic rocks of the Ritter Range pendant are a composite of two entirely different aged volcanic arcs- one in the Jurassic and the other in the Cretaceous (Fig. 96, Maps 33-35). The eastern side of the pendant exposes Paleozoic metamorphosed sedimentary rocks that are related to those of the Mount Morrison pendant to the south at Convict Lake. There is a lot of geological history in the Ritter Range pendant, but the trail only crosses the older part of the Jurassic volcanic rocks. The rocks were deformed during the development of the Cretaceous volcanic arc and again throughout the intrusion of the Late Cretaceous plutons. Textures in the rocks, such as foliation, cleavage, veins, and small-scale folding, were partially caused by the Sierra Crest shear zone system. Before this strike-slip fault, the meta-volcanic rocks were deformed by thrust faults (Tobisch and others, 2000), causing the rotation of bedding to a near vertical position. This event also worked on the volcanic rocks of the Mount Morrison and Goddard pendants.

The Cretaceous part of the pendant is made of thick caldera collapse deposits (Kistler and Swanson, 1981; Fiske and Tobisch, 1994) that are of similar size to the modern Long Valley caldera. The volcanic breccias compose the Clyde Minarets,

Mount Ritter, and Mount Banner. Similarly, the northernmost part of the pendant preserves a Triassic collapse caldera at Mount Dana, which was mainly filled by metamorphosed volcanic breccias, tuffs, and lesser arkose and polylithic conglomerate.

The Ritter Range pendant contains a wide variety of units; for the section of trail between Shadow Lake and Thousand Island Lake, the guidebook uses the numbered subdivision listed below. The rock is complicated enough that while writing this book I had hiked across these units twice, and I still was not comfortable with identifying all of the rock types. To complicate things, different publications by the same authors use varying map unit names and show the contacts and units slightly different between studies. The below listed rock units I present to show the composition of the layers, all of which are metamorphosed. Even though Map 34 is complicated it is still a simplification due to scale limitations. More recent work has identified bed-parallel faults and complex deformation in the section (Sharp and others, 2000, Tobisch and others, 2000). The most important thing to know while hiking this section of the trail is the volcanic rocks of the Ritter Range pendant are highly layered, and commonly the beds are laterally continuous for several kilometers. The pendant is rich in interesting structural features so that just about anywhere one wanders there are things to examine, such as minor folds, cleavage, en echelon veins, deformed volcanic clasts, minor faults, etc.

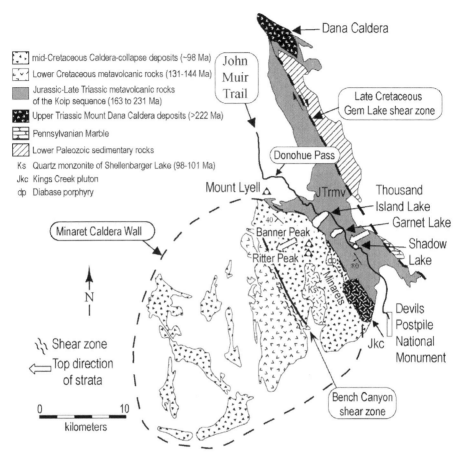

Figure 96. Geology of the Ritter Range pendant. The Cretaceous caldera outline from Fiske and Tobisch (1994). The eastern side of the pendant, through which the John Muir Trail traverses, is composed of Jurassic metamorphosed volcanic rocks that are probably related to those in the Mount Morrison and Goddard pendants. Age ranges summarized from Kistler and Swanson (1981), Schweickert and Lahren (1993), Fiske and Tobisch (1994), and Tobisch and others (2000).

Rock units as mapped by Tobisch and others (1977), Map 34

2   Gently dipping felsic volcanic rocks

3   Bedded to massive lithic lapilli tuffs, fine slaty tuffs, local tuff-breccia, marble, accretionary lapilli tuff, and minor ash-flow tuffs. Unit 3 is medium gray and lighter gray when weathered, and it has green epidote pods and veins. The layering in the rock varies in thickness, from laminated to very thick. Some of the units contain accretionary lapilli, which are spherical accretions of tuff material, possibly formed by volcanic ash settled during a rainstorm so that the water droplets agglomerated the ash particles together.

4   Ash-flow tuff and crystal tuff. The rock is medium gray, forming very thick sheets/beds, and contains abundant flattened pumice clasts (possibly equal to unit JTrx of Huber and Rinehart 1965).

5   Chaotic slump deposits consisting of massive lithic lapilli tuff, fine tuff and tuff-breccias matrix, including chunks of bedded fine tuff, accretionary lapilli tuff, and porphyritic sills.

6   Intrusions of fine-grained to aphanitic sills and dikes- some are locally porphyritic, and ranges from mafic to felsic in chemical composition (maybe unit Kibc of Tobisch and others, 2000).

7   Fragmental intrusion of a sill, locally flow banded, generally appears similar to massive lapilli tuff and tuff breccia (and maybe deformed tuff).

8   Lava flows and/or sills and lava breccia, mafic in composition; includes minor amounts of lapilli tuffs and fine-grained slaty tuffs. Unit 8 is very dark as compared to the other volcanic units. It is marked by small phenocrysts in a very fine-grained matrix.

9   Tuff breccia, lapilli tuffs and fine-grained slaty tuff; bedded to massive.

The trail drops approximately 244 m (800') on the northward descent of the steep ridge to Shadow Lake (**Shadow Lake access point**). Along this section, the trail possibly switchbacks over units 3 and 4, but the area is heavily forested so this is uncertain. The trail then recrosses through units 4 and 3, possibly unit 9, then over units 3, 8, 3, 8, 3, 8, 4, and 7 during the traverse around the south side of Shadow Lake. Most of the outcrops are discontinuous and tree covered, adding to the difficult of distinguishing between the various alternating rock types. The trail stays in unit 3 and then unit 7 along Shadow Creek, and then ascends over unit 4.

The trail crosses over unit 7 and then 4, 8, and 3, and climbs to a saddle. Descending to Garnet Lake (Map 34), the trail intersects with units 3, 8, 3, and 8 at the outlet. If all these numbers seem a bit confusing, it is because the rock types in the layered volcanics alternate, are relatively thin, and the JMT recrosses the same beds while switch backing. Along the north side of Garnet Lake, metamorphosed tuffs have large, deformed spheroids that are possibly accretionary lapilli or spherulites (Fig. 97).

The trail ascends from Garnet Lake, crossing tuffs and sills (unit 8), runs through a saddle, and then passes Ruby Lake on the east side. It crosses over a lava flow or sill unit just north of the lake (the rocks are metamorphosed, so the exact origin for the tabular igneous bodies is difficult to determine) and then over undifferentiated volcanic rocks (JTru). The JTru is an U.S. Geological Survey designation, mapped by Huber and Rinehart (1965). Tobisch and others (1977) did not extend their detailed map this far north.

The east side of Ruby Lake (Map 34) has a small lateral moraine. In the southwest cliffs above the lake are dark gray and dark reddish layering in the rock. The reddish layers were folded into a synform, a U-shape pointing more or less upwards. This observation is important because it indicates the thickness of the meta-volcanic units was changed during deformation.

The metamorphosed tuff units to the east of Ruby Lake have abundant thin white quartz veins that record left-lateral strike slip deformation in the rock. The veins define en echelon patterns, are folded when viewed in the horizontal exposures (Fig. 98), and in vertical exposures some are boudinaged (Fig. 99). Phenocrysts have asymmetric recrystallized tails that indicate right-lateral shear. This pattern is the exact same structural style as described for the deformed aplite dikes on the west margin of the Mount Morrison pendant. Both these areas were deformed by the Sierra Crest shear zone system (Green and Schweickert, 1995), however, these features have not been studied in detail at these locations. It is significant that the tuff near Ruby Lake contains the same type of deformation markers as the Gem Lake shear zone in the northeast part of the pendant (Greene, 1995). Tobisch and others (1977) described a large amount of strain and deformation in the Ritter Range pendant, which appears to be a composite of several events. The left-lateral sense-of-slip markers as recorded by the quartz veins were overlooked in this study. Recently Sharp and others (2000) described similar fabrics to the east of Shadow Lake, and provide geochronologic data that indicates the deformation was synchronous with the Gem Lake shear zone.

Figure 97. Metamorphosed tuff in the Ritter Range pendant exposed along the north shore of Garnet Lake. Oval shapes may be deformed accretionary lapilli, spherulites, or concretions. Swiss army knife in lower right corner for scale.

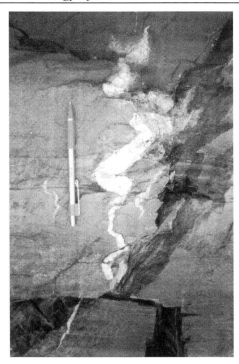

Figure 98. Folded quartz vein in metamorphosed tuff east of Ruby Lake. Outcrop surface is horizontal. The asymmetrical folds indicate left-lateral shear.

The trail passes Emerald Lake on the west side, and then descends to Thousand Island Lake (Map 35), which occupies the glacially carved valley. Along the south shore of Thousand Island Lake, near the outlet, are accessible outcrops of metamorphosed tuff and sedimentary rock. The contact between the lighter colored tuff and the dark-reddish meta-sedimentary rock has two important features allowing the top direction of the formations to be determined. One, the tuff includes abundant rounded fragments decreasing in size to the west, suggesting it is a basal surge deposit at the bottom of the tuff. Surge deposits are formed by a blast of volcanic material moving over the ground. The second important observation is the tuff contains ripped-up pieces of the underlying gray siltstone, indicating the tuff is younger. Layering in the meta-sedimentary rock is isoclinally folded, meaning that the angle between the straight fold limbs is less than 10

Figure 99. Boudinaged quartz vein in a near vertical exposure. This location is two meters from the folded vein shown in Figure 98, and is part of the same vein set. The boudins plunge 41 degrees to the southwest. A horizontal cut through this outcrop would probably delimit asymmetric folds. The simultaneous vertical stretching and horizontal shortening with a component of shear is a type of deformation referred to as transpressional. Hand lens for scale.

degrees. The steeply tilted formations are younger towards the west, consistent with the overall map pattern of the Jurassic-Triassic volcanic rocks along the trail being overlaid by less deformed Cretaceous caldera deposits to the west.

Over-looking Thousand Island Lake are the spectacular Ritter Peak (4,010 m, 13,157') and Banner Peak (3,945.4 m, 12,945'), which are both composed of metamorphosed Cretaceous caldera fill deposits belonging to the Minaret Caldera (Tobisch and Fiske, 1994). Calderas are large volcanic craters rimmed by cliffs and at the center have relatively flat, down dropped floors. The inferred location of the caldera rim is shown in Figure 96. Because calderas define depressions they act as traps containing volcanic breccias, lava flows, ash flow tuffs, and landslide blocks. Also, they commonly erupt domes fields that cut through the caldera fill material. The plutons in the southern part of the pendant intruded this volcanic pile (the volcanic rocks were probably erupted from the earlier plutons) suggesting the intrusion depth was relatively shallow, about 2 to 3 km below the surface.

The ascent to the north of Thousand Island Lake traverses over the Kuna Crest Granodiorite (Kkc). These exposures are southern isolated lobes of the Kuna Crest Granodiorite and are composed of more felsic, or lighter-colored, rock than the main mass of granodiorite to the north. It should be noted that this smaller body was not dated, and the correlation by Huber and Rinehart (1965) may be incorrect.

The trail passes over the following steeply tilted meta-volcanic map units: near the lake outlet is JTrss (composes the stream crossing), and then JTru, JTrc, JTru, JTrx, Kkc, and finally JTru (Huber and Rinehart, 1965). Recall that JTr stands for Jurassic-Triassic age. The lowercase c in JTrc stands for calcareous sedimentary rocks, including thin-bedded tuffaceous sandstone, siltstone, marble, conglomerate, and slate. This unit also contains fossils of pectinoid pelecypods (similar to clams) of Early Jurassic age (Huber and Rinehart, 1965). JTru refers to undifferentiated Jurassic-Triassic volcanic rocks mainly consisting of crystal lithic tuffs, and tuff breccias (Fig. 100). JTrx is composed of crystal tuffs. The view southward over Thousand Island Lake provides clear examination of the thick, continuous volcanic layering dipping close to vertical (Fig. 100).

The Kuna Crest Granodiorite is the oldest unit of the Tuolumne Intrusive Suite (Fig. 102). The suite consists of four main intrusions with the younger ones intruded into the center of the older plutons. Each younger intrusion is progressively more enriched in quartz than the previous outer ones.

Figure 100. Metamorphosed lithic-rich tuff. Note the abundant small angular fragments and foliation. Lens cap for scale.

Figure 101. Overview of Thousand Island Lake, looking south, showing the near vertical dipping beds in volcanic rocks of the Ritter Range pendant.

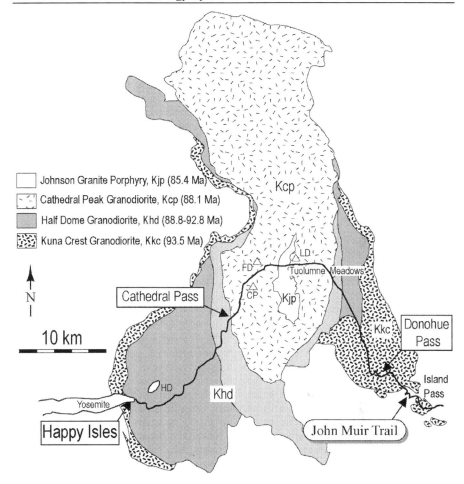

Figure 102. Generalized geology of the Tuolumne Intrusive Suite. LD= Lembert Dome, FD= Fairview Dome, CP= Cathedral Peak, and HD= Half Dome. The geology was modified from Bateman and Chappell (1979), Bateman and others (1983), and Huber and others (1989). See appendix II for the sources of the age data.

Kuna Crest Granodiorite (Kkc)

The Kuna Crest Granodiorite, Maps 34-37, dated at 93.1-93.5 Ma (Coleman and others, 2004) and probably closer to 93 Ma from cross cutting relationships with the Half Dome Granodiorite (Coleman and others, 2002, 2004), is dark colored, fine- to medium-grained rock containing abundant biotite and hornblende (Fig. 102). This rock unit is one of the oldest in the Tuolumne Intrusive Suite. Foliations in the pluton are vertical and striking W-NW. This rock is best exposed at Donohue Pass. The unit has been correlated to other outlying intrusions lining the outer part of the Tuolumne Intrusive Suite, including the Glen Aulin tonalite (also called the quartz diorite of May Lake), the Glacier Point tonalite, and the Grayling Lake tonalite (Bateman and Chappell, 1979, Huber and others, 1989). For simplification, I have included all these granitoid bodies as part of the Kuna Crest Granodiorite. Even so, it should be noted that compositionally these units are variable. A comparison of rock textures for the major units of the Tuolumne Intrusive Suite is shown in Figure 103.

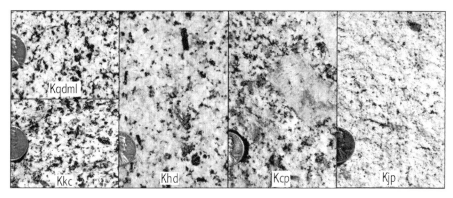

Figure 103. Rock textures of the Tuolumne Intrusive Suite. KqdML = quartz diorite of May Lake and Kkc = Kuna Crest Granodiorite, both are part of the outermost concentric shell surrounding the entire suite. Note the more or less equigranular texture. Khd = Half Dome Granodiorite, note the euhedral large phenocrysts of hornblende. Kcp = Cathedral Peak Granodiorite, overall rock color is slightly lighter than the Half Dome Granodiorite, and includes prominent megacrysts of potassium feldspar. Kjp = Johnson Granite Porphyry, a very fine-grained, equigranular rock of overall very light color. Scale is approximately the same for each photograph.

The trail leaves the southern isolated body of Kuna Crest Granodiorite (Kkc?) to cross over a small hornblende diorite intrusion (Kdis). Nearby are two, shallow, small lakes to either side of the trail. The diorite is coarse-grained and has abundant large crystals of black hornblende (Fig. 104). Near the contact between the diorite and the Kuna Crest Granodiorite, pieces of the diorite are included (surrounded by) the Kuna Crest Granodiorite. This indicates the diorite is older than the Kuna Crest Granodiorite. Furthermore, the diorite is not deformed, indicating that its age of intrusion was younger than the deformation in the meta-volcanic rocks of the Ritter Range pendant.

The trail goes over undifferentiated metamorphosed volcanic rocks (JTru) to gain Island Pass (waypoint 306693E, 4178701N), which is composed of the Kuna Crest Granodiorite (Kkc). On the north side of Island Pass (Map 35) the trail crosses a large inclusion of meta-volcanic rock (JTru) surrounded by Kkc. Before the trail reaches Rush Creek it enters a northern spur of the Ritter Range pendant, which is composed of metamorphosed Jurassic to Cretaceous volcanic rocks of various types (JTru). Several thin screens and blocks of JTru are crossed by the JMT during the descent from Island Pass.

Near the trail junction for Davis Lake (Maps 35) is a contact with a sheared medium-grained diorite (KJd) intruded into the meta-volcanic rocks of the Ritter Range pendant. In the more coarse-grained zones, anastomosing foliation (from shearing) is well exposed. Anastomosing foliation is a wavy layering in the crystal fabric of the rock, produced by deformation at appreciable pressure and heat. Later, more brittle, shear zones produced a series of right-stepping cracks, partially lined by epidote, as seen in the horizontal exposures (this pattern is called en echelon, see Fig. 105). En echelon fractures and faults are a common feature of youthful fault zones. As deformation proceeds, the cracks become linked so that a fault surface of greater continuous length is formed. This pattern operates at all scales, from micro-cracks in a rock to the development of major plate boundaries such as the San Andreas fault.

After passing the northern spur of the Ritter Range pendant the trail returns to the Kuna Crest Granodiorite (Kkc). From the Rush Creek crossing the trail stays in the Kuna Crest Granodiorite (Kkc) on the climb up to Donohue Pass.

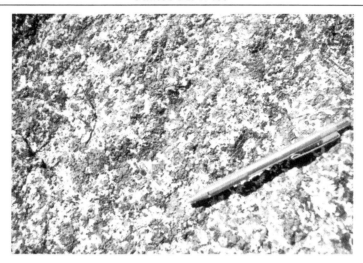

Figure 104. Close up photograph showing texture of hornblende diorite unit (Kdis) at Island Pass. The large dark crystals are hornblende.

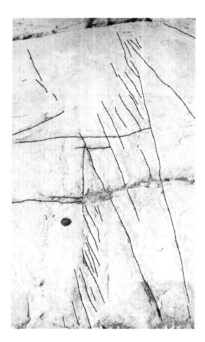

Figure 105. Outcrop photograph of en echelon fractures in a near horizontal glacial carved slab. The cracks are right stepping, indicating left-lateral shear. The fractures are marked by a halo of hydrothermal alteration, bleaching the granitic rock white. Lens cap for scale.

Donohue Pass (3,368 m, 11,050', Map 36, waypoint 301999E, 4181473N)

East of the pass lies Donohue Peak (3,664.4 m, 12,023'), and to the west is Lyell Peak (3,997 m, 13,114'). The view south of Donohue Pass from left to right includes: Koip Crest, June Mountain (the ski runs are visible from the pass), Benton Range, White Mountains, Mammoth Mountain, Red Slate Mountain, and the summits of Ritter and Banner. On Koip Crest is a highly contrasting lighter-colored aplite sill cutting the Kuna Crest Granodiorite. The view north from the pass down Lyell Canyon has Mount Conness directly inline with the valley in the distance. This peak is in the Tuolumne Intrusive Suite. Donohue Pass is in biotite-rich Kuna Crest Granodiorite. The amount of biotite in the intrusion decreases inward towards the contact with the Half Dome Granodiorite. Lyell Canyon is about 16-km long (10 miles) glacial-carved valley that drops only 518 m (1,700 ft) in elevation after the initial steep descent from the pass. Numerous lateral and recessional moraines are found lining the valley floor, remnants from the now almost completely retreated Lyell glacier. Mount Lyell, named after Charles Lyell, an eighteenth century geologist from England, is the source of the glacier. At the base of the northern side of the peak lies the last remnant of the active Lyell glacier, first discovered and described by John Muir in 1873. When John Muir first studied the glacier, the toe of the glacier probably extended close to the JMT. John Muir first noted evidence of active glaciation in the Lyell Fork by identifying glacial rock flour or fine silt in the flowing stream, possibly close to the ford along the JMT. Even now the streambed at the Lyell Fork crossing may have a little bit of light green-gray sediments from glacial outwash!

The trail descends from Donohue Pass to cross the Lyell Fork at the north end of a tarn, then it descends and crosses to the east side of the stream. The trail recrosses to the west side via a bridge at the 2,995 m (9,825') elevation. The steep descent into Lyell Canyon is mostly on Tioga stage till, however, the trail crosses a few small, glaciated, outcrops of the Kuna Crest Granodiorite.

Halfway down Lyell Canyon the trail intersects the contact between the Kuna Crest Granodiorite and the Half Dome Granodiorite. Unfortunately, the contacts are covered by till, meadow (Qal), and forest.

Half Dome Granodiorite (Khd)

The Half Dome Granodiorite, Maps 37-44, with U-Pb dates ranging from 88.8 Ma to 92.8 Ma (Coleman and Glazner, 1997; Coleman and others, 2002, 2004), was described by Calkins (1985) as "Medium gray, medium-grained; relatively uniform in color and texture. Well formed crystals of biotite and hornblende are common [especially large ones at Half Dome]. Plagioclase predominates over potassium feldspar." Bateman (1992) described the Half Dome Granodiorite as consisting of a Megacrystic facies (Khdp) and an Equigranular facies (Khde). Megacrysts are large phenocrysts, and equigranular refers to a rock in which all the minerals have a similar grain size. The Equigranular facies accounts for most of the pluton. The Half Dome Granodiorite contains numerous mafic inclusions of diorite, and schlieren composed of euhedral hornblende and sphene, and interstitial quartz. The Half Dome Granodiorite has a range of age data showing formation over two million years by multiple pulses of possibly dike-like smaller intrusions (Coleman and others, 2002; Glazner and others, 2002). The Half Dome Granodiorite cross cuts the tonalite of Glen Aulin, dated at 93.1 Ma (Coleman and others, 2002, 2004), and the Kuna Crest Granodiorite to the east, dated at about 91 Ma but from the above new age constraints is probably closer to 93 Ma.

The pluton composes about 40% of the exposed surface areas of the Tuolumne Intrusive Suite (Fig. 102). Large parts of the Half Dome Granodiorite may appear overall uniform, but increasing subtle differences are being noted in the unit. In the glacial carved slabs northeast of Olmstead Point, the Half Dome Granodiorite contains a variety of different mafic inclusions and complex steeply dipping layering. Locally, schlieren or layered zones where heavy, mafic minerals, such as hornblende and sphene, were concentrated clearly show the heterogeneous character of the granodiorite. Finally, the mechanical characteristics of the Half Dome Granodiorite produces massive-cored domes that develop excellent sheeting on their exteriors.

An additional 3.2 km (2 miles) down the canyon, as mapped by Bateman and others (1983), the trail leaves the equigranular facies of the Half Dome Granodiorite (Khde) and goes into the porphyritic facies (Khdp, contact waypoint ~298675E, 4191139N). Farther down the canyon, the contact between the Half Dome and Cathedral Peak Granodiorites is exposed in glacially polished slabs just west of the trail (waypoint 297535E, 4192732N). The gradational contact between the two plutons is about 1.5 km (~0.9 mile) into the canyon from Tuolumne Meadows trailhead. One must watch for the subtle color change in the rock slabs; the Half Dome Granodiorite is grayer in color, and characterized by large crystals of hornblende and biotite, plus mafic inclusions, which the Cathedral Peak Granodiorite does not have. The slabs just west of the trail show many subtle internal contacts where concentrations of mafic minerals lie along planar features (Fig. 106), and locally one ladder dike was observed at 297529E, 4192744N (Fig. 107).

Cathedral Peak Granodiorite (Kcp)

The Cathedral Peak Granodiorite, Maps 39-42, 86 Ma (Bateman, 1992) and 88.1 (Coleman and Glazner, 1997; Coleman and others, 2004), is composed potassium feldspar, quartz, plagioclase, and minor biotite. The potassium feldspars form large crystals, some up to 12 centimeters in length, called **megacrysts**. This is the most distinctive feature of the rock. Bateman and Chappell (1979) reported the size of these extremely large potassium feldspar phenocrysts to decreases away from the contact with the Half Dome Granodiorite, and at the same time the total amount of potassium feldspar remains constant at about 20%. The distribution of the megacrysts is heterogeneous, almost forming zones completely absent of megacrysts and zones clustered with megacrysts. Numerous aplite dikes may be found in the pluton, attesting to the presence of a late stage aqueous phase (high H_2O content) throughout the final crystallization of the pluton. The dikes range from being completely barren of potassium feldspar megacrysts to those containing a high percentage crystals. Locally, discontinuous pegmatites of potassium feldspar and quartz are also present, some occurring as pockets in otherwise fine-grained dikes of aplite. Piccoli and others (1997) mapped subtle zonations of minerals and

> textures in the aplite dikes, interpreting the dike material to be locally derived melt crystallized late in the pluton's history. The Cathedral Peak Granodiorite comprises the largest of the plutonic units in the Tuolumne Intrusive Suite (Fig. 102). Foliations are generally vertical and striking to the W-NW.

Scattered throughout the flat meadows of Lyell Canyon are glacial polished slabs of the Cathedral Peak Granodiorite (Kcp). Commonly, the low hills along the canyon floor are lined by Tioga till (Qg). Several large glacial erratics, close to the size of cars, are scattered throughout the canyon. When passing such erratics, look at the rock composition to see which pluton they originated from.

The trail descends through a low-walled corridor cut into the Cathedral Peak Granodiorite that has weathered out large megacrysts of potassium feldspar. Also, several thin subhorizontal sills (tabular intrusions) of aplite are in the outcrops. To form sills, the pressure of the molten rock had to exceed the lithostatic pressure (the weight of all the rock above the sill) so space could be made for the molten material. This pressure is immense when you consider that several kilometers of rock were above the sills.

The trail passes by a small white alkali flat (some times pond) on the west side, crosses a wooden footbridge over Rafferty Creek, and at the junction for the Vogelsang trail the JMT heads west.

This segment of the guide ends at the trail junction for the Tuolumne Meadows trailhead (waypoint 294789E, 4193588N), which lies just southeast of Lembert Dome- a prominent feature towards which the northbound hiker can not fail to observe. Lembert Dome is a large roche moutonnée, a ridge of rock cut by glaciers. If one scrambles up to the top of the dome, numerous glacial erratics can be found lying on top of this formation (Fig. 38), indicating that glaciers once covered Lembert Dome. A short distance north from the JMT is the Tuolumne Lodge and pack station (where hot showers can be purchased from noon to 3:00 pm for non-guests as of 2007), and also nearby is a ranger station with a public phone. One kilometer farther along the JMT, a junction is reached leading to the Tuolumne store, cafe, and post office. Also, here the Pacific Crest Trail departs from the John Muir Trail. The Pacific Crest Trail continues northwestward near the crest of the Sierra Nevada to link up with the Cascade volcanic arc and follows these volcanoes to the Canadian border.

From the trail junction, the JMT heads westward and the next section of the guidebook covers the remaining portion of the Tuolumne Intrusive Suite and glacial features of the Yosemite region.

Figure 106. Photograph of banding of mafic phenocrysts within the gradational contact zone between the Half Dome and Cathedral Peak Granodiorite.

Figure 107. Photograph a ladder dike exposed in a horizontal glacial slab. GPS for scale. Not the layering within the dike runs somewhat perpendicular to the wall rock contacts, and the orientation of the k-spar megacrysts are lying parallel with the mafic mineral banding.

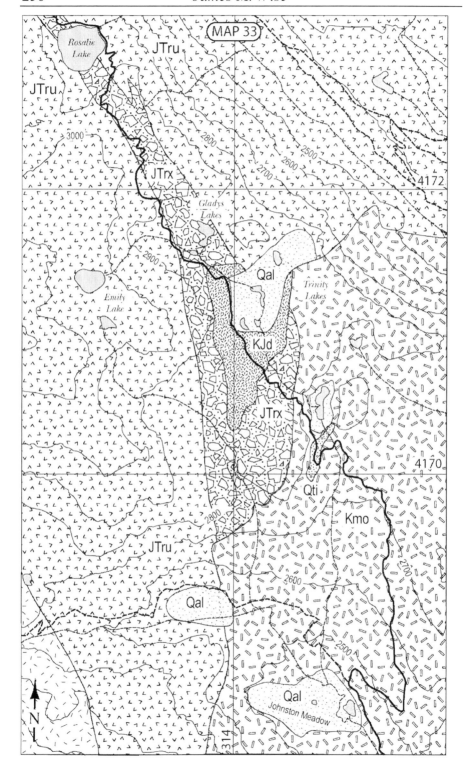

Geology of the John Muir trail 257

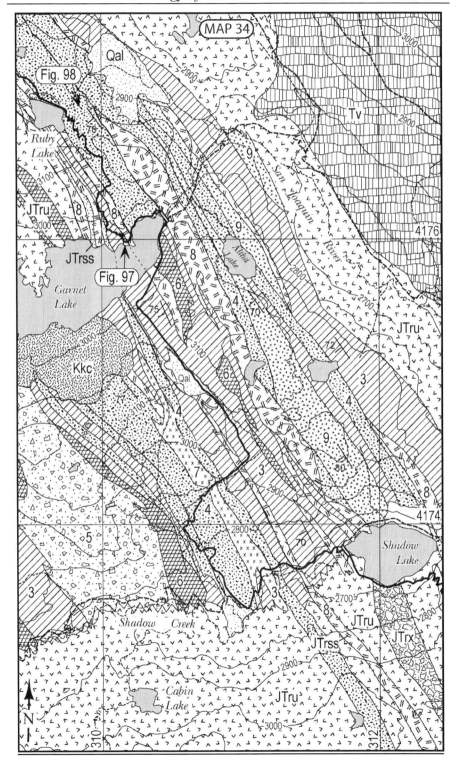

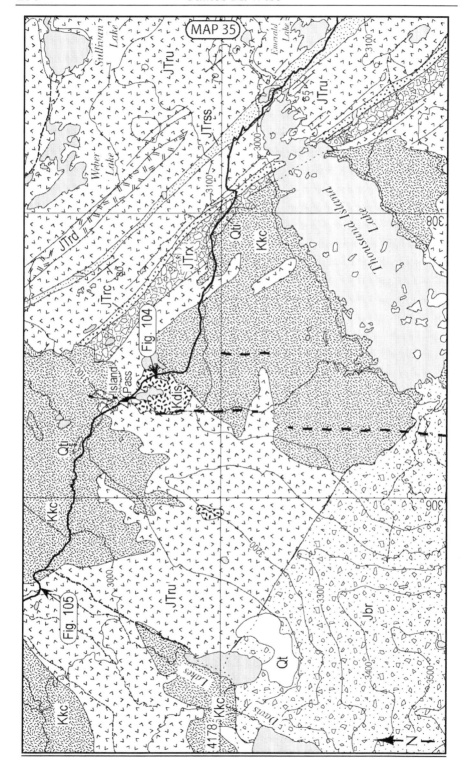

Geology of the John Muir trail 259

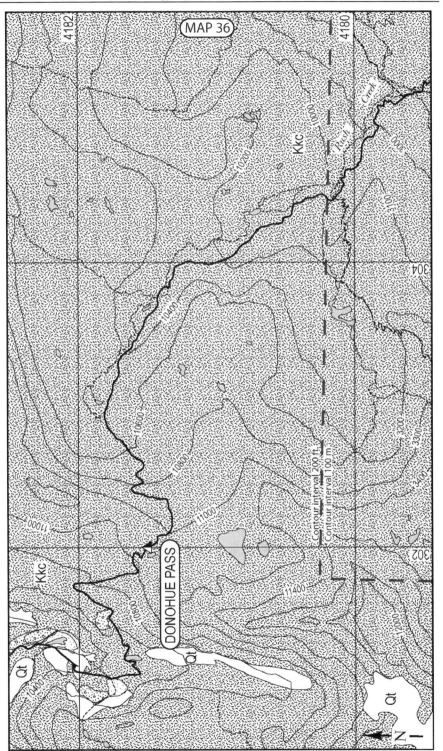

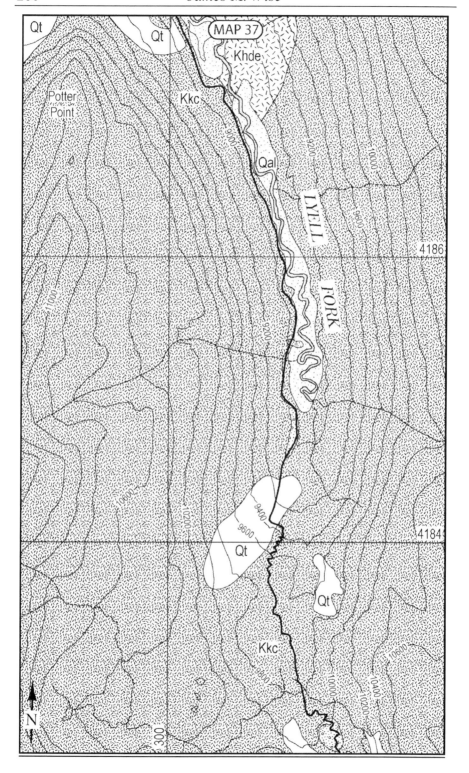

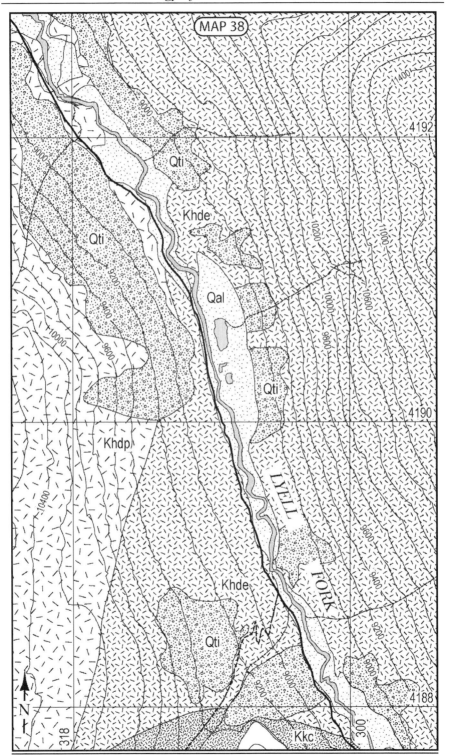

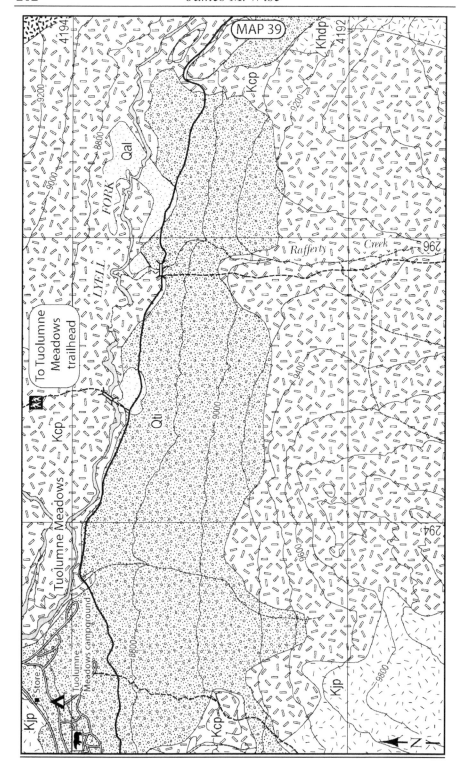

CHAPTER 10

TUOLUMNE MEADOWS TO YOSEMITE

Access: Happy Isles, Cathedral Lake, and Tuolumne Meadows trailheads.
Distance: 35.2 km (21.9 miles).
Maps: 39-44.

A Short History of Glacial Research in Yosemite

In 1869 John Muir wrote of Yosemite Valley "...*a grand page of mountain manuscript that I would gladly give my life to be able to read. How vast it seems, how short human life when we happen to think of it, and how little we may learn, however hard we try!*" Yet John Muir went on to describe many aspects of the glacial history of the Yosemite region. In his time, the extent to which glaciation was responsible for the formation of the valley was a centerpiece debate amongst geologists. Joseph Whitney and Clarence King, both eminent geologists of the time, held the opinion that Yosemite resulted from a catastrophic event, which pulled apart the valley and down-dropped a large block represented by the valley floor, producing the vertical valley walls. Clarence King did record some observations regarding glacial evidence around Yosemite Valley. In 1873, Joseph LeConte, assisted by John Muir, documented several lines of evidence for the glacial origin of the valley. In addition, John Muir put forth the idea of the glaciers eroding preferentially along the regional joint pattern in the igneous rocks, which he summarized as "...*the grain of the rock determines its surface forms.*" Muir took glaciation too far, attributing all of the landscape development to the forces of ice. Numerous workers to follow proposed hypotheses on how the glaciers carved the valley, and to what extent ice versus water erosion was responsible.

In 1930, an U.S. Geological Survey geologist named Francois Matthes produced a comprehensive study of the glacial features of Yosemite, documenting three stages of glaciation. The terminology used by Matthes conflicted with work along the eastern side of the Sierra by Blackwelder published in 1931. Blackwelder's terminology has become standard in the Sierra Nevada, for reasons discussed in the beginning of this guide under glacial geology. Matthes' 1930 publication is a Sierra Nevada classic filled with important observations, and the beautiful illustrations, many of which have become entrenched in books describing the geology of Yosemite Valley. Unfortunately, some of these ideas still used today are wrong. For example, Matthes used the popular geomorphic approach at the early part of the 1900's, called the Cycle of Erosion, proposed by William Davis, a professor at Harvard University, to describe the development of landforms through incremental uplift and then weathering down of the terrain during protracted periods of stability. Matthes applied the Davis geomorphic model to the formation of Yosemite while not having knowledge on the nature of uplift or relief development for the Sierra Nevada.

Over the last three decades, glacial studies in the Sierra Nevada have focused on relative dating techniques of till on the east escarpment, drill-testing sediments at Owens Lake, and stratigraphic studies of sedimentary formations along the western foothills. Landform analysis is taking advantage of digital elevation models to characterize entire watersheds, develop river profiles along all the drainages, and compare along range variations using detailed statistics. The geomorphic evolution of the Sierra Nevada is now considered a continuum of nearly ongoing uplift and accompanied with repeating or alternating periods of glacier and stream erosion. Landforms represent a series of different time interval equilibriums where the shape of the land adjusts to the forces acting upon it. The details in understanding the balance or response between forces and the resulting landscape are complex. The approach was first explored by Grove Gilbert (1877), and then later further developed by Strahler (1950) and Hack (1960), which has become a field of science called process geomorphology (Sack, 1992). This still growing branch of science has yet to be directly applied to the formation of Yosemite Valley.

Along the Tuolumne to Yosemite segment of the trail, some of the glacial features described by Matthes are passed on the descent to culminate in one of the world's preeminent glacial sculptures of Yosemite Valley. Even so, areas remain to be mapped with greater detail, such as the moraine deposits surrounding Little Yosemite Valley.

Trail description

This segment of the John Muir Trail begins at the Tuolumne trail junction (waypoint 294789E, 4193588N), just south of Highway 120. It is best reached by hiking south from the Tuolumne Meadows trailhead. An alternative starting point is located at the Cathedral Lakes trailhead, yet this access point misses a rock unit called the Johnson Granite Porphyry, which is the youngest unit of the Tuolumne Intrusive Suite. In this segment of the JMT the western half of the Tuolumne Intrusive Suite is traversed, and beautiful glacially carved features are observed during the descent into Yosemite Valley. The trail crosses over the Johnson Granite Porphyry, the Cathedral Peak Granodiorite, and the Half Dome Granodiorite, plus till deposits from the Tioga and Tahoe stages of glaciation.

From the Tuolumne Meadows trailhead junction, the JMT heads west, passing several low slabs of the Cathedral Peak Granodiorite. A trail junction for the east side of the Tuolumne Meadows campground, post office, store, cafe, and public telephones is soon reached. The trail continues to the west over Tioga till (Qg) to a trail junction for Elizabeth Lake and the Tuolumne Meadows Campground (waypoint 292915E, 4193670N). The trail stays on Qg, passing over the covered contact between Cathedral Peak Granodiorite (Kcp) and Johnson Granite Porphyry (Kjp) about half way to the next trail junction. Nearby Unicorn Creek, a trail junction leads to the west side of the Tuolumne Meadows Campground. The route ascends a short hill west of the tributary to Unicorn Creek, passing the only exposures of the Johnson Granite Porphyry that is close to the JMT. The best outcrops are located in low relief slabs 60 m south of the trail in which a sub-horizontal sill composed of predominately k-spar megacrysts is exposed. A sill is a tabular, or sheet-like-body of igneous material cutting through older rock.

> **Johnson Granite Porphyry (Kjp)**
>
> The Johnson Granite Porphyry, Maps 39-40, with zircon dates ranging from about 82 to 85.4 Ma (Stern and others, 1981; Fleck and others, 1996; Coleman and Glazner, 1997; Coleman and others, 2004), is the youngest plutonic unit of the Tuolumne Intrusive Suite. It intruded the Cathedral Peak Granodiorite, forming a small body in the center of the intrusive suite (Fig. 102). It is composed of an equigranular matrix of quartz and feldspar imbedded by larger phenocrysts of euhedral (well formed), six sided, gray quartz. It includes some megacrysts of potassium feldspar, but no where near as abundant as the Cathedral Peak Granodiorite. Its porphyritic texture and lack of systematic internal fabrics suggests it crystallized deep in a volcanic neck (Bateman and Chappell, 1979; Titus and others, 2002). The rock has small miarolitic cavities, open spaces from high vapor pressure in the rock while crystallizing, which suggest the intrusive formed within 4 km of the paleosurface. The contacts of the Johnson Granite Porphyry with the Cathedral Peak Granodiorite are sharp. The contacts along the JMT are unfortunately covered in till. Flinn and Reid (1986) suggested the Johnson Granite Porphyry co-existed with the Cathedral Peak Granodiorite as two hybrid melts (molten rock compositionally separated from the same source) and had limited mixing between the magmas. However, the radiometrically determined age reported by Fleck and others (1996) places the intrusion 5 million years after formation of the Cathedral Peak Granodiorite.

From the sill outcrop, the trail contours for a short distance westward over forested till deposits, then passes some more exposures of Kjp, and then returns to Qg. After the junction leading to the Tuolumne Visitor Center (just to the north and down the hill, waypoint 291585E, 4193870N), and farther on to Soda Springs/Parsons Lodge/Glen Aulen, the trail makes a short climb up a hill composed of Qg. It passes some slabs of Kcp at the hilltop, gently descends over Qg to Budd Creek, crosses a small wooden bridge, and then comes to the **Cathedral Lakes trailhead junction (ENTRY POINT, Map 40, waypoint 290444E, 4194120N)**. The Cathedral

Lakes trailhead is along Highway 120, several hundred meters north of the JMT.

MULTIPLE-PULSE FORMATION OF THE TUOLUMNE INTRUSIVE SUITE

The series of concentric shells in the Tuolumne Intrusive Suite, along with their composition and internal chemical and mineralogic variations, led Bateman and Chappell (1979) to conclude that this giant composite plutonic suite developed by a series of pulse injections or movements of magma. In their model, the outermost more mafic plutons, here grouped into the Kuna Crest Granodiorite, made an earlier large magma chamber that commenced crystallizing into a solid from the outer margins towards its center (Fig. 108A). Before the chamber was completely solidified the inner still molten core of the pluton received an immense influx of new magma, the Half Dome Granodiorite, which increased the size of the magma chamber, both disrupting and cutting across the earlier crystallized margin of the Kuna Crest chamber (Fig. 108B and C). The Half Dome Granodiorite has age data ranging from 90.7 to about 92.6 Ma and complex internal features suggesting prolonged multiple intrusions (Coleman and others, 2002). A new, more felsic, resurgence of magma into the chamber was repeated throughout the intrusion of the 88.1 Ma Cathedral Peak Granodiorite (Coleman and Glazner, 1997), which nearly doubled the size of the pluton (Fig. 108D). Finally, the Johnson Granite Porphyry was emplaced at a slightly later time, perhaps using the same conduit or feeder structure as the earlier granodiorites, but it may have been intruded into a completely solidified composite pluton. This last intrusion at the core of the Tuolumne Intrusive Suite suggests perhaps a higher level of emplacement in the crust as based on rock textures that may indicate the magma fed eruptions to the surface.

The model by Bateman and Chappell (1979) is based solely on patterns in the Tuolumne Intrusive Suite and does not provide supporting evidence from the metamorphic screens and roof pendants along its margins. The western margin of the suite intruded the already solidified approximately ~102-105 Ma Taft and El Capitan Granites and 94.6 Ma Sentinel Granodiorite (Calkins and others, 1985; Coleman an Glazner, 1997; Ratajeski and others, 2001) of the Yosemite Intrusive Suite. To the northeast, the Tuolumne Intrusive Suite cuts rocks of the northern Ritter Range pendant. How far did these older rocks extend into the present position of the Tuolumne Intrusive Suite? Several aspects of the emplacement model may have to be modified, including the possible contribution from tectonic opening to provide space for the magma. The Sierra Crest Shear zone, the Gem Lake segment, may continue along the eastern margin of the Tuolumne Intrusive Suite, and has been called the Cascade Lake shear zone (Davis and others, 1995). This fault may have provided a control on the NW-elongate form of the Cathedral Peak Granodiorite, and possible deformed the older marginal shells of the Half Dome and Kuna Crest Granodiorite. Furthermore, the locus of magmatism may have originally been controlled by the now consumed Mojave-Snow Lake fault.

Likewise, the earlier part of the emplacement history of the Kuna Crest Granodiorite may have to be examined in greater detail. Along the west side of the suite, the elongate metamorphic rock in the May Lake screen contains intensely developed penetrative deformation fabrics that indicate subhorizontal right-lateral shearing. These structures have not been directly dated and remain unknown if they formed during the emplacement of the Kuna Crest Granodiorite. The quartz diorite at May Lake also presents some potential complications where it is in contact with the Half Dome Granodiorite. The foliated quartz diorite, defined by mineral orientation and flattened mafic inclusions, is truncated by compositionally banded granodiorite. This contact relation suggests the initial Kuna Crest magma chamber was at least locally deformed before emplacement of the Half Dome Granodiorite, perhaps by inflation of the magma chamber. Subsequent foliations developed throughout the remainder of the Tuolumne Intrusive Suite do not define a simple pattern paralleling the major rock contacts. Instead, foliation is locally perpendicular to the internal contacts, suggesting that the fabric formed by deformation instead of lateral crystallization or by magma movement. New geochronologic and field data on the Tuolumne Intrusive Suite is demonstrating more complex, multi-stage intrusion, perhaps in part as a dike complex (Coleman and others, 2002; Glazner and others, 2002). Although, I have not seen much evidence for dike formation of the suite and many aspects of this hypothesis remain controversial.

The model shown in Figure 106 perhaps should be revised so that the Kuna Crest magma chamber starts off smaller and then is expanded with each subsequent intrusion. Progressive expansion of the magma chamber may be from tectonic opening along strike-slip faults because rocks of the Saddlebag Lake pendant show no evidence for forceful emplacement or ballooning of the plutons. There is no data on the initial or original size of the Kuna Crest magma chamber, making any reconstruction speculative.

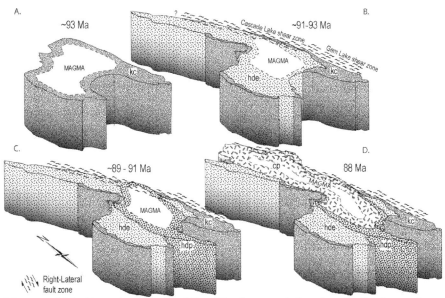

Figure 108. Development of the Tuolumne Intrusive Suite by pulses of magma introduction (modified from Bateman and Chappell, 1979). Initial intrusion of the Kuna Crest Granodiorite (kc) was followed by injection of the equigranular mass of the Half Dome Granodiorite (hde), possibly accompanied by a component of right-lateral strike-slip faulting along the eastern margin. Several other pulses of magma may have been introduced during the formation of the Half Dome

Granodiorite, but the inner porphyritic facies (hdp) makes a distinct large unit in the suite. Emplacement of the Cathedral Peak Granodiorite (cp) made a very elongate body, and then finally was intruded by the Johnson Granite Porphyry (not shown).

Figure 109. View westward over Tuolumne Meadows towards Cathedral Peak, which defines the high point on the skyline. The landforms are a result of glacial erosion followed by sheeting of the granitic rock, and the flat-valley floor is made by a combination of flooding and migrations of the river channel position. Photograph was taken from the top of Lembert Dome.

From the Cathedral Lakes trailhead junction (waypoint 290444E, 4194120N), the trail heads southwestward, climbing up hummocky deposits of Tioga till. Higher up, Cathedral Peak (3,325.5 m, 10,911') comes into view on the east side of the valley (Fig. 109). This glacial horn has steep cliffs controlled by exfoliation along vertical joints. Cathedral Peak formed a rock island or arete high above the ice throughout the Tioga stage glaciation. To the south of Cathedral Peak is a broad U-shaped pass through which ice flowed towards the west. As the trail crosses Cathedral Creek (waypoint 288594E, 4193078N) Fairview Dome is directly to the north (Map 40, Fig. 18). The dome shape of rock formations around Tuolumne Meadows is from glaciation and exfoliation by sheeting. The top and southern slopes of Fairview Dome has large k-spars weathering out with positive relief, suggesting that the upper part of the dome was not covered by Tioga stage glaciers. A short distance to the north along the same ridge, the smaller Marmot Dome has well developed glacial polish up to its summit. Broken exfoliation slabs high on the north face of the dome have been polished, indicating that the sheeting predates the Tioga stage glaciation.

The trail descends and contours around Cathedral Lakes basin along the east side. The lakes are large glacial-carved depressions in the Cathedral Peak Granodiorite. Eventually the lake will become filled by sediment, forming a flat meadow and perhaps will become forested.

Cathedral Pass (2,950 m, 9,680', Map 41, waypoint ~287549E, 4190073N)

The trail executes a short climb up to Cathedral Pass where it crosses a small patch of Tioga till and alluvial deposits (Qal). Views from the pass are rather subdued as compared to other passes along the JMT. Just beyond the pass the JMT enters a northeast-trending valley lined by Tioga till.

The trail ascends over Tioga till, descends along the east side of an arete named Columbia Finger, and then continues down over Qg to the east side of Long Meadow.

Midway in Long Meadow are slabs along the east side of the trail in which k-spar megacrysts of the Cathedral Peak Granodiorite compose up to 80% of the rock and reach up to 15 cm in length (Fig. 110). These protruding crystals characterize the type of holds and rock texture taken advantage of by rock climbers in the Tuolumne Meadows area. To develop larger-sized crystals the nucleation rate for phenocryst growth must be low, limiting potassium feldspar crystallization to the megacrysts. Nucleation of a crystal is the starting point of a crystal. A melt that cools with abundant nucleation sites will result in numerous small crystals. Moreover, to form megacrysts the crystallization rate must be slower, meaning the magma cooled slowly. If a melt is cooled rapidly it quenches, forming either a very fine-grained rock or a glass. The volcanic flows at Mono Craters is a good example of quickly cooled magma. This outcrop of megacryst rich Cathedral Peak Granodiorite is at the opposite extreme for cooling history when compared to the Mono Craters rhyolite flows.

The trail next heads south, crosses the drainage to the west side and follows it through scenic Long Meadow. At the south end of Long Meadow, exposures of the Half Dome Granodiorite are located east of Sunrise High Sierra camp. The contact between the Cathedral Peak Granodiorite and the Half Dome Granodiorite is exposed on a ridge just north of the trail (waypoint 286154E, 4185849N). The contact is sharp, marked by a change from a hornblende granodiorite to a potassium feldspar megacryst granodiorite. The contact continues to the south beneath the Tioga till valley fill of Long Meadow. Looking farther south at the north face of an unnamed dome, the rock is cut by a two-meter thick aplite dike. Beginning at the outcrops on the south side of the trail that has several thin vertical aplite dikes (3 cm to 13 cm thick), the first exposures Half Dome Granodiorite here is on the west side of the Tuolumne Intrusive Suite.

The trail descends from the meadows near Sunrise High Sierra camp (usually mosquito infested), at first contouring, then losing elevation rapidly. At the base of the slope the trail crosses an internal contact in the Half Dome Granodiorite (Khd- the contact is between two textural units inside the Half Dome Granodiorite). Next, it traverses parallel along the crests of two lateral moraines.

The trail descends over a small patch of older Tahoe till (Qta), crosses a valley filled with Tioga till (Qg), and passes Moraine Dome along the north side (map 43). On the southern side of Moraine Dome are well-formed lateral moraines from which the dome is named, however, these are not visible from the trail. These moraines indicate that the dome was not covered in ice from the Tioga stage glaciation. The weathered surface of Moraine Dome also indicates that it was not glaciated since it was covered by ice from the earlier Tahoe stage glacier (Matthes, 1930). On top of the dome, I found patches of Qta till containing boulders of the Cathedral Peak Granodiorite with weathered out k-spar crystals. On the basis of similar rock weathering where aplite dikes or crystals of feldspar protrude from the surrounding rock, Matthes (1930) determined that neither Half Dome nor the summit of Clouds Rest were under ice during either of the last two glacial stages.

From Moraine Dome, the trail descends along heavily forested lateral moraines. At the junction for the trail to Half Dome (waypoint 278775E, 4180306N), the JMT heads to the south. From the junction it is 4.2 km (2.6 miles) to Half Dome, a detour worth the time and effort because steel cables on the east side of the dome provide access to the most spectacular view of Yosemite Valley.

Geology of the John Muir trail

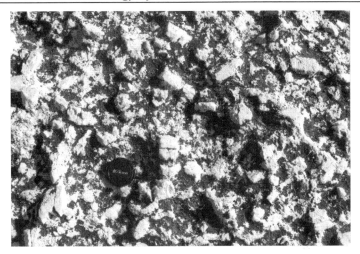

Figure 110. Photograph of potassium feldspar megacrysts (light colored rectangles) in the Cathedral Peak Granodiorite weathering out with positive relief. The crystals are up to 5 cm in length. Lens cap for scale.

Formation of Half Dome

The main controlling factors in the formation of Half Dome are the erosion of Tenaya Canyon, the thickness of glacial ice, and exfoliation controlled by northeast-striking vertical joints that parallel the northwest face (Fig. 111). The northwest face does not have any glacial polish on the wall. The northwest base of Half Dome has a marked change to a lower slope angle, marking the position of the more recent glacial trim lines. The steep face of Half Dome has retreated to the northeast away from this trim line by mass wasting of panels of exfoliated rock. Glacial erosion at the base stimulates vertical panels of rock to be exfoliated and transported away on the glacier. Sketches of glacial ice scouring away the face of Half Dome, while dramatic, perhaps only happened in the most severe and older ice ages, if at all. Glacial ice from the Tahoe and Tioga stages certainly did not abrade the face of Half Dome. Likewise, the rounded south face of the dome has numerous deep rills or troughs that probably indicate glacial ice did not abrade the surface after the Sherwin stage. The top of Half Dome has numerous bathtub sized or larger solution pockets, a feature that only develops with prolonged periods of exposure. The round form of the dome is from sheeting, a process related to the released pressure on the granodiorite. When the granodiorite first crystallized it was probably at 6-8 km or greater depth. Now it does not have the weight of the overlying rock on it, so the surface undergoes expansion, forming onionskin like sheets of rock. The sheeting of Half Dome is near horizontal at the summit and then curves down the face to the north of the prow to continue in partially preserved patches of near vertical exfoliation flakes (Fig. 112). Northeast-striking vertically sheeted panels of rock are stacked to either side of the main northwest face. Exfoliation is substantially controlled by topography, and therefore the formation of Half Dome is dependent on the development of the adjacent canyons.

The first real map oriented and detailed description of the glacial geology seen along this segment of the John Muir Trail was by Matthes in 1930. At this time, Matthes had to survey all the topography to be used as a base, and did not have the advantage of aerial photographs to help record the location of moraine deposits. Recently, Schaffer (1997) did a bulky publication on the glacial history of the Yosemite. He points out that there are numerous inconsistencies in the early reconnaissance style work of Matthes, but commits errors himself by presenting interpretations as data and by lacking a coherent or systematic approach to glacial geology beyond that of paranoid bashing of the U.S. Geological Survey geologists. So while the work of Matthes is dated, it is the recommended starting point for learning about the glacial history of Yosemite. As first outlined in this segment of the guide, the glacial history of Yosemite has had a rich and emotionally heated history starting with Whitney and Muir. It is interesting to see more recent works commit to the

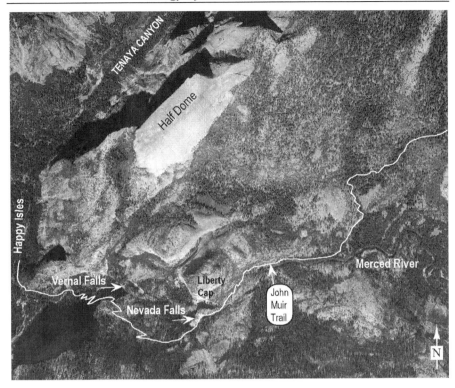

Figure 111. Aerial photograph showing the region of Half Dome and the end of the John Muir Trail. Photo was originally at about 1:20,000 scale taken by the U.S. Forest Service in 1944. Darkest areas are shadows, mottled dark patches are pine stands, and lightest areas are barren granitic domes.

same pitfalls from over 100 years ago. Perhaps it is the stunning setting of Yosemite that prohibits researchers from maintaining objective work. For the hiker completing the John Muir Trail from south to north, the glacial features of Yosemite should be familiar, now just at a much larger scale.

The JMT heads downhill from the Half Dome trail junction (Map 43), crossing over multiple crests of lateral moraine material deposited by the Tioga stage glaciation, to descend into Little Yosemite Valley. The flat-floored valley is covered by a mixture of Tioga till, Holocene alluvium, and fluvial deposits from the Merced River. Several recessional moraines are present here, including one that the trail descends onto when first reaching the valley floor. A flat sandy hike west along the south side of a lateral moraine brings one to a trail junction.

Figure 112. Photograph of the upper portion of the northwest face of Half Dome, showing exfoliation sheeting wrapping around from the flat top part of the dome to near vertical sheets below the prow. All surfaces are a result of exfoliation.

The trail leads westward over flat ground along the north bank of the Merced River. To the north is a thick Tioga stage lateral moraine that partially covers the eastern ridge of Liberty Cap, which is the prominent dome overlooking the trail.

At the trail junction above Nevada Falls (waypoint 277088E, 4178233N), the JMT takes the south fork and crosses a wood bridge over the Merced River just upstream from the top of the falls. A few thick aplite dikes are crossed by the trail. The other fork is the Mist trail, which makes a steep descent over stone stair steps, along the cliffs next to Vernal Falls. The Mist trail is shorter and provides a steeper descent. Near the bridge crossing the Merced River at the top of Nevada Falls a short trail leads to the brink allowing one to look down from the lip of the waterfall.

The Giant Staircase- glacial steps

While standing on the bridge at the top of Nevada Falls, which drops 181 m (594 ft.). Try to imagine the Merced Glacier in the Tioga stage glaciation forming an ice fall or crevasse field as it moved down the cliff, as first suggested by

LeConte in 1873. The Merced River longitudinal profile defines a pronounced staircase pattern, a feature present along most glaciated valleys (Fig. 113). Glacial staircase profiles (also called glacial steps, rock steps or riegels) were also passed along the JMT in LeConte Canyon and upper Evolution Valley, to name just two examples. The Merced River staircase is perhaps the most pronounced in the Sierra Nevada. Huber (1981b) described the erosion of the Tuolumne River, showing a stepped longitudinal profile between Hetchy Hetchy and Mt. Lyell. Similar to Vernal Falls, the drop at Mist Falls along Woods Creek, Kings Canyon, also makes a near vertical step in glaciated terrain.

Steps in river profiles, called nick points, may also develop in nonglaciated terrains where combination of interlayered weak and more resistant layers interact with stream erosion. A well-known example of a nick point involving sedimentary formations is Niagara Falls. Along the Merced River the bedrock is uniformly composed of the Half Dome Granodiorite, making mechanical breakdown more dependent upon joint directions and spacing. During erosion each nick point, or cliff, moves upstream, or as Matthes (1930) described, recedes headward by quarrying of the bedrock. Such nick points may also become semi-fixed upon areas of more resistant, less jointed, bedrock. Huber (1981b) hypothesized that temporary canyon lining of volcanic deposits may propagate nick points into the bedrock that persist after removal of the channel armoring material. Such an explanation by itself probably does not account for the vertical nick points at Nevada and Mist Falls because there are no volcanic rocks perched on the canyon walls.

Detailed mapping of joints at Vernal and Nevada Falls has never been done, so it remains unknown if these waterfalls lie upon more erosion resistant rock. Nevada Falls at least appears to have a northeast trend to the cliff face, suggesting that the nick point may be controlled by peeling back of slabs of rock, which are bounded by NE-striking joints. Note at the top of Nevada Falls the amount of river incision or erosion into the brink of the cliff and then compare this to the form of Vernal Falls where essentially no down cutting has occurred. At the base of Vernal Falls amphitheater is a post-glacial talus apron formed by exfoliation of the over steepened walls. The talus cones line both canyon walls all the way down to Happy Isles. Liberty Cap may have been poking above the Tioga stage glacier, but was completely covered by ice throughout the Tahoe stage glaciation (Matthes, 1930). Now a hand full of large, rounded erratics are strewn about on the glacial valley floor, a very different Little Yosemite Valley than 20,000 years ago. Glaciation will most likely occur in the future, reforming all of Yosemite Valley. It is worth re-emphasizing that the Giants Staircase formed by multiple periods of glaciation and intervening periods of stream erosion interacting with heterogeneous bedrock.

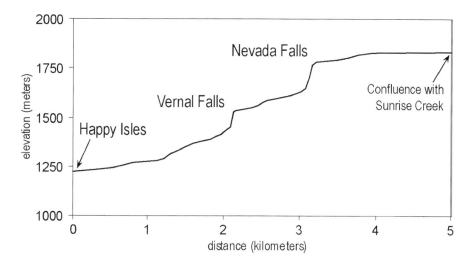

Figure 113. Longitudinal profile of the Merced River showing staircase topography. Matthes (1930) called the lower sloped sections between nick points treads. Note the difference between the vertical and horizontal scales.

After crossing the Nevada Falls bridge, the John Muir Trail makes a short climb and then comes to a trail junction for Glacier Point (waypoint 276683E, 4177851N). The JMT takes the right or west fork and then descends to another trail junction at Clark Point (275819E, 4178148N); the trail going north leads back to the Mist Trail. The JMT continues westward to descend many switchbacks along the north side of the high Panorama Cliffs. At the base of the talus slope the lower junction with the Mist Trail near the Merced River is reached (waypoint 275432E, 7178280N). After passing the Mist trail junction, the trail drops to the bridge at the Merced River below Vernal Falls. From the bridge there is a good view of Vernal Falls, which is 97 m (317 ft) in height. In the Spring it is a roaring torrent and often by Autumn it is reduced to a mere trickle. Note that the base of Mist Falls is lined by talus, material that probably represents post-glacial exfoliation of the cliff. Near the bridge are a drinking fountain, bathroom, emergency phone, and Yosemite tourists.

After crossing the bridge to the north side of the Merced River, a short climb leads to the final down hill and the asphalt paved section of the trail. Part of the trail was wiped out by the 1990 rock avalanche deposits triggered by an earthquake near Lee Vining. The trail rounds a promontory called Sierra Point where the Half Dome Granodiorite is well exposed in the blasted rock of the trail. The trail descends over moss covered talus slopes and groves of California Live Oak. At the bottom of the hill the John Muir Trail ends at Happy Isles (waypoint 274633E, 4178845N), elevation 1,234.4 m (4,050 ft), on the east side of the grandest glacially carved valley of Yosemite (Map 44). The final rock type along the JMT is the Half Dome Granodiorite. This unit composes much of Glacier Point, Washington Column, and Royal Arches in Yosemite Valley.

A wood sign is posted at the end of the trail, giving the mileage of the trail, and beyond this a bridge foundation (the bridge across the Merced River was destroyed in the 1997 flood). Now at the end of the JMT, the vertical relief between the summit of Mount Whitney and the Yosemite Valley floor is 3,183 m (10,444 feet)- yes, the journey was all down hill!

Near the Happy Isles trailhead, on July 10, 1996, a large block of rock, approximately 500 feet across, fell 1,800' from near the top of the Glacier Point wall to the southwest of the Merced River. The impact of the rock sent a shock wave of air that knocked over numerous trees, and sent a dust cloud 3,000 feet into the air. Unfortunately, eleven people were injured and one person was killed by the knocked over trees. Some of the trees were chopped off several meters above the ground and the toppled portions point away from the impact point (Wieczorek and others, 2000). This type of geologic event is common in Yosemite Valley, and anywhere else in the Sierra Nevada along cliffs where the forces of gravity, ice, water, and time will eventually tear down the highlands. Other recent large rock falls in Yosemite include the 1982 Cookie Cliff slide that blocked highway 140 in the lower part of the valley, and the March 10, 1987 Middle Brother rockslide. A series of blocks have been peeling off from the Glacier Point cliff above Camp Curry, having three major rock falls from 1998 to 1999. It is worth noting that not all the geological processes are slow. The sudden transport of large volumes of rock is known as mass wasting.

The terminus of the JMT, or if southward bound, the start, is where a sharp change in the river gradient forces the Merced River to slow down, thereby decreasing the energy for transporting sand, gravel, and boulders along the river bed. Happy Isles is a confined alluvial fan walled in by the canyon cliffs. Northward of Happy Isles the crest of a possible recessional moraine is preserved. It is near the old union of the Merced and Tenaya glaciers, which has a medial moraine pointing westward.

A short walk north of the Happy Isles trailhead leads to the free shuttle bus, which completes a user-friendly loop through Yosemite Valley. Yosemite is one the most magnificent geologic exposures in the world. It is worth a visit to the Yosemite Visitor Center to see the scale model of the valley showing the main igneous rock types, an useful introduction to the geology (recently removed from display, and hopefully it will be upgraded in the Visitor center remodel). For those more interested in the specifics of the geology, I recommend the geologic map by Calkins (1985), usually for sale at the Visitor Center. The bedrock of Yosemite Valley is composed of over eight plutonic rock types, for which their geologic history with respect to the Tuolumne Intrusive Suite is poorly understood.

It is also worthwhile to spend sometime looking at the several compositions of plutonic rock in the wall of El Capitan, which has to be one of North America's largest continuous vertical outcrops, standing at about 884 m (2,900') high with most of the wall being vertical to slightly over vertical. Half Dome, though having a higher summit than El Capitan, has a slightly less than vertical wall of 549 m (1,800'). In California, only Tehipite Dome, a 975 m (3,200') high wall to the north of Kings Canyon, beats El Capitan in height but not in steepness. Some world class big walls include: Polar Sun Spire, a 1,433 m (4,700') high wall at Baffin Island; the 1,524 m (5,000') east face of Trango Towers, Pakistan; the 1,524 m (5,000') east face of Cerro Torre, Patagonia in Argentina; and the 1,067 m (3,500') Troll Wall of Norway. All of the above walls are in glaciated landscapes. El Capitan is climbed more often because of its temperate climate, easy approach, proximity to large population centers, and unsurpassed quality of rock.

As a final section of this guide, I present several different aspects on the geology of Yosemite in the next chapter, focusing on the nature of the valley fill, and the plutonic history.

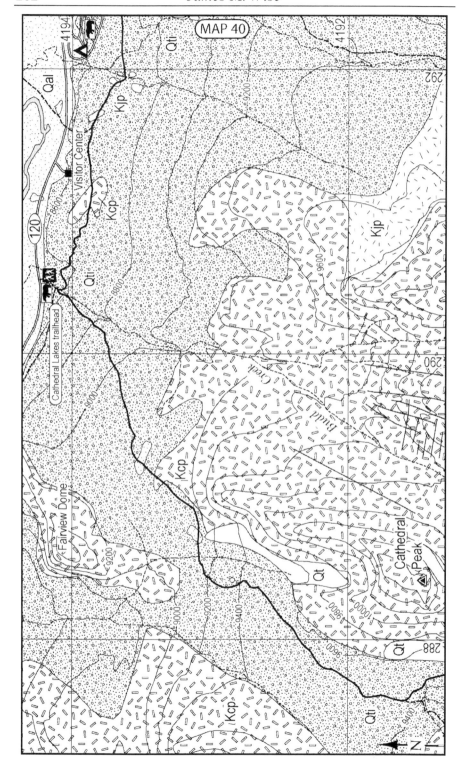

Geology of the John Muir trail 283

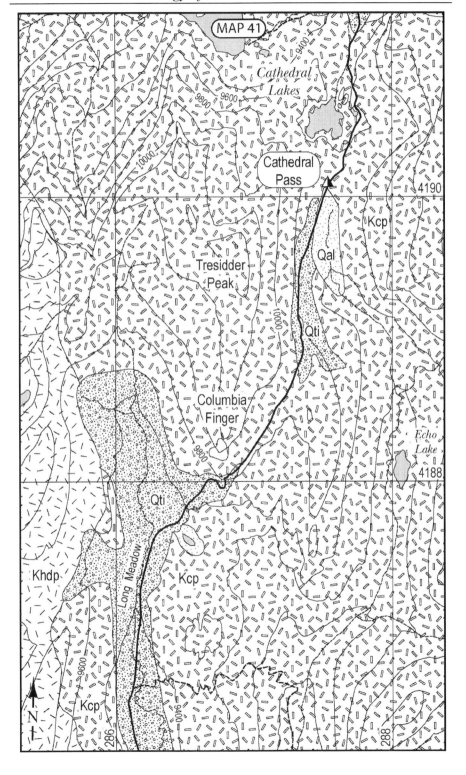

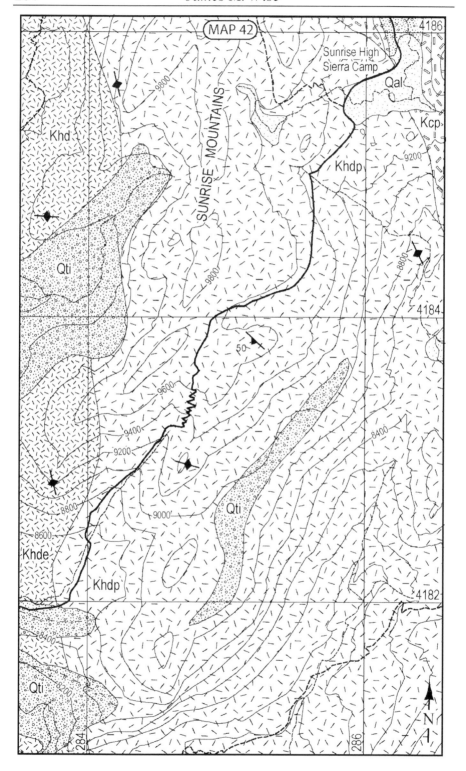

Geology of the John Muir trail

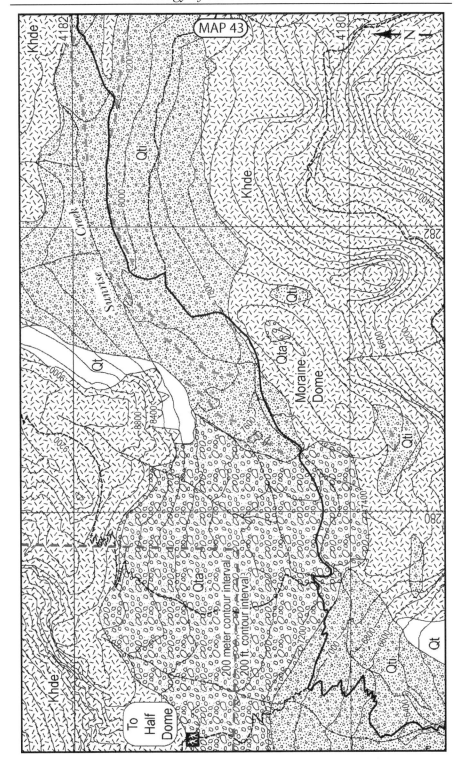

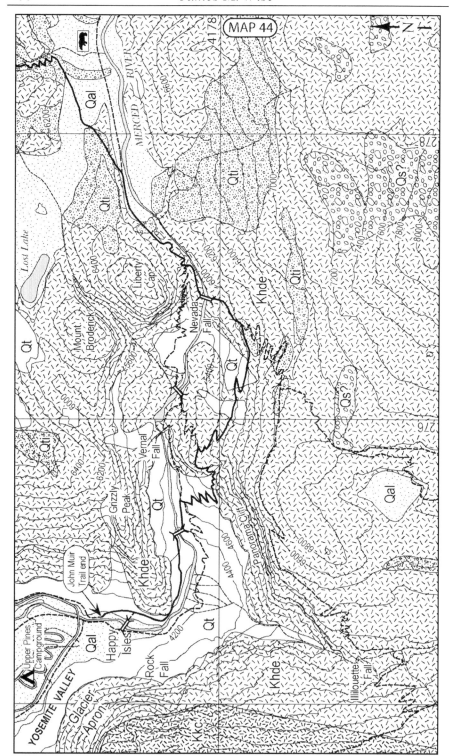

CHAPTER 11

YOSEMITE VALLEY

Those having just completed John Muir Trail will probably be more interested in finding the showers, the pizza stand, and the bar. However, if you find yourself waiting for a bus out of the valley, or are diehard on geology, the below information should provide for an in-depth experience of Yosemite Valley.

A wide variety of publications cover the geology of Yosemite Valley. I recommend the beautifully illustrated book by King Huber titled *The Geologic Story of Yosemite National Park*. Anyone wanting to take a serious examination of the geology should acquire a copy of the 1985 U.S.G.S. geological map of the valley by Calkins (generally available in the visitor center). Instead of repeating many of the classic Yosemite Valley geology examples, I use this final section of the guide to address a few specific problems and detail some interesting geologic features that are generally not in the literature.

JOINT CONTROL OF YOSEMITE VALLEY

The principle control on the geometry of rock formations of Yosemite Valley is imparted by the pattern of joints in the bedrock, for which Huber (1989) gives several fine examples. Joints provide the weakness along which erosion performs, and in the case of glaciation the interaction results in spectacular landforms developed parallel to the prominent joints. From the northwest face of Half Dome to Cathedral Spires, joints define the orientation of the cliffs (Fig. 114). In general, erosion interacting with the regional joints produces many spectacular landscapes, including the deep canyons at Zion National Park. Study of jointing at Yosemite can be used to understand the overall large-scale form of the valley and the small-scale surface features observable at the outcrop.

For any one large body of rock there is typically a minimum of three directions of joints that break the rock mass into various sized blocks. At the scale of the entire valley there are several primary orientations of joints that are NE, NW, and nearly NS striking. NE-striking joints control the pattern of Half Dome, Tenaya Canyon, the gorge between Mount Broderick and Liberty Cap, the main steep faces of the Three Brothers, the ravines and spires at Cathedral Rocks, and the deep gorge on the west side of El Capitan. Also, the wide-spaced NW-striking joints define the headwall at Taft point, but the nearby Fissures are along close-spaced NE-striking joints (Fig. 115). The near NS-striking joints focused erosion at the deep Indian Canyon to the west of Royal Arches. Most books about the formation of Yosemite Valley show the Three Brothers as more or less the type example of joint control on the morphology of the cliffs. The regional joints also directed the overall zigzag geometry of the valley that is obvious in Figure 114. The NE-striking joint set, such as in the trend of Tenaya Canyon, is the dominant control on the shape of Yosemite Valley. Note that the age of the NE-striking joint set is unknown. However, the NE-striking joints, many with minor left-lateral fault slip, are part of a regional pattern of fractures that extends southward to Mount Whitney (Lockwood and Moore, 1979).

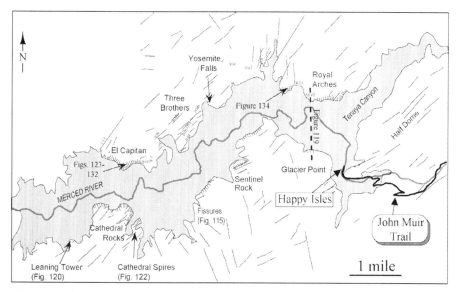

Figure 114. Map of the main joint pattern of Yosemite Valley, which defines a preferred NE-striking direction (thin lines) that controls the form of the valley and major rock formations. Joints were located by use of aerial photographs, topographic maps, and field observations. Locations of subsequent figures are marked. The gray area represents the valley fill material, and hatches mark the cliffs.

The older joints partially control the location modern rock fall and exfoliation along the high cliffs. Exfoliation in Yosemite has had few studies, but the research by Bahat and others (1999) records interesting observations on El Capitan where there is little, if any, interaction with the older joint sets. Through laser surveys of the scars left behind by spalled-off exfoliation flakes, which delimit large areas of concoidal or fan-shaped fracture surfaces left on the rock face (Fig. 116), the pattern was related to crack propagation in a stress field where the maximum direction of compression is vertical. These fractures are related to the unbounded or unsupported lateral flanks of the steep valley walls and represent a different type of fracture formation than the earlier formed systematic joint sets that probably were controlled by the orientation of stress during tectonism. Upon close examination, most of El Capitan is scarred by concoidal fracture surfaces. The El Capitan Granite seems particularly suited to develop this type of fracture pattern because of the granite's high quartz content and equigranular medium-grained rock texture. Concoidal fracture surface morphology can also be found on rock formation near Half Dome (Fig. 117). Many exfoliation flakes still plastered to the wall of the valley are in the process of buckling and collapse. Some are delicate enough that on certain rock climbing routes they are referred to as *expanding*, meaning that under loads a climber may place the exfoliation flake it will flex and open. These disconcerting flakes may also be called hollow, named after the reverberating sound they produce when struck.

The walls of Yosemite Valley have yet to be mapped for the amount of area covered by glacial polish versus surfaces carved by exfoliation. It would be interesting to see where the highest level of preserved glacial polish is at for the entire valley, which would perhaps yield information on the thickness of the latest Tioga stage glacier in Yosemite. Yet more likely such a map would reveal or emphasize the extent to which exfoliation has shaped the valley walls.

Some have proposed that deep weathering of the Sierra Nevada block controlled the dominant landforms. Most joint sets of the High Sierra today show no sign of oxidation of the mafic minerals. On the other hand, farther down slope to the west in the foothills, reddish soil profiles are well developed and the joints certainly provide the focus of weathering. Stripping of this soil profile or regolith is suggested to determine the location of deeper canyons. In contrast, I suspect that major canyons would have followed the same predominate joint sets with or without deep weathering of the bedrock.

Figure 115. Photograph looking east at very closely spaced NE-striking joints that control the Fissures near Taft Point. View is towards the north-northeast.

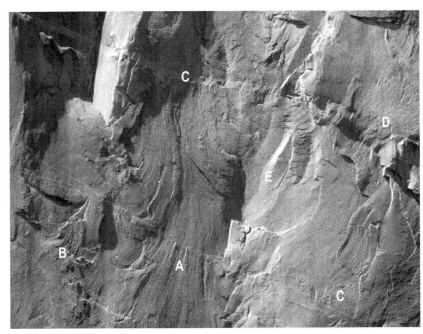

Figure 116. Photograph of fracture surface morphology on the nose of El Capitan. Vertical field of view in the photograph is approximated 200 feet (see Fig. 124 for location). With oblique lighting on Yosemite walls there is an incredible amount of rough detail to the rock that was formed by the process of exfoliation cracks. Essentially 100% of the surface area in this photograph was formed by exfoliation cracks. **A.** Bottom center of photograph has downward sweeping pattern of lines on the rock that is known as plumose structure, a feature that forms during the growth of a crack. **B.** Left side of photo has stair-stepped thick appearing flakes that are commonly formed near the end of a propagating crack, a feature called hackles. **C.** Thin lighter-colored planar features running diagonal from lower right to upper left of photograph are aplite dikes. The hackles and plumose structures vary in direction and spacing where crossing the aplite dikes. **D.** In upper right corner of the photograph note how plumose structure in part initiates on the edge of the dike, but likewise crosses it. **E.** Roughly developed concentric bends in fracture surface are known as rib structures.

Figure 117. Photograph of fracture surface morphology exposed in cliffs below Half Dome. Hackles create the staircase breaks at several scales, representing en echelon branches developed at the leading edge of the exfoliation crack. Plumose structure is also nicely expressed in the upper left center portion of the photograph, showing the fracture grew from left to right. Note the trees on the upper right skyline for scale.

Mega-sheeting at Royal Arches

At Royal Arches the process of exfoliation is working to remove the angular glacial trim line where the valley wall changes from steep to low angle (Fig. 118). The arches are immense sheets breaking through the side of the valley. Note how the arches terminate against the regional high-angle joint set, which defines the slot-like gully that sets Washington Column aside from the rest of the valley wall. Royal Arches perhaps represent the largest arches in the Sierra Nevada. One consequence of these thick onion-skinned style slabs of granite is the house-sized blocks lining the talus apron beneath the arches (approached from the Ahwahnee Hotel parking lot).

Figure 118. Photograph of extremely large exfoliation sheets developed at Royal Arches.

Yosemite Valley Fill

The flat-floored Yosemite Valley was produced by the combination of small terminal moraines overlain by subsequent fluvial flood plain deposition. These deposits are interfingered with talus and rock fall deposits that become volumetrically more significant closer to the base of the cliffs. Near El Capitan are series of terminal moraines from the Tioga stage, but from here to Happy Isles the exposed valley floor is dominated by fluvial deposits. A detailed description of glaciation in the Sierra Nevada was presented in Chapter 1. Glacial deposits within Yosemite Valley has not yielded much data on the timing of the major ice ages. Meandering of the Merced River had effectively reworked or eroded away any surface topography of the recessional moraines.

Seismic Reflection Data

The sediment fill of the Yosemite Valley averages 366 m (1,200') in thickness, and is deepest in the center of the valley and tapers towards the canyon walls. The bedrock profile, as known from a seismic survey conducted in 1935, has deepest portion of the valley between the Ahwahnee Hotel and Camp Curry where it is nearly 610 m (2,000 feet) thickness of valley fill (Gutenberg and others, 1956). This data set indicates three main thick members to the valley fill (Fig. 119). Most importantly, the bulk of Yosemite Valley was carved out by the older stages of glaciation. Both the Tahoe and Tioga stages did not further deepen the valley, instead these glaciers rested on the older glacier deposits in Yosemite Valley. We are probably well overdue for a new 3D seismic survey of the valley because additional details can certainly be imaged with today's technology.

Drill Hole Data

Two water supply wells drilled in the early 1970's near Yosemite Creek give a poorly described geologic log of the sediment types filling the valley. Both of the holes are about 1,000-feet deep. Cuttings from these holes reveal three major stratigraphic units in the subsurface, mainly composed of bouldery sand and only minor clay. Unfortunately, the lithologic logs for these drill holes are vague, having not been recorded by geologists. Nonetheless, they do not appear to represent filling of a Yosemite Valley Lake (e.g., Fig 75 of Huber, 1989), because such deposits tend to be well sorted, meaning the grains in the deposits are all of similar size. Matthes (1930) argued that gravel and sand from the still active glaciers up canyon filled Lake Yosemite. This filling would proceed by a combination of depositional processes, including delta progradation, turbidite deposition, and fall out of suspended particles. These depositional types probably would not lead to what appears very poorly sorted or variable grain sizes found in the drill holes. The sedimentary fill in Yosemite Valley has valuable information about several major ice ages and this record has yet to be examined in detail. A few exploration drill holes in Yosemite could significantly contribute to our understanding of the ice ages, and the total amount of time over which the valley developed. The idea of Lake Yosemite remains controversial and will not be resolved until either high-resolution seismic surveys and drill core samples of the section are obtained.

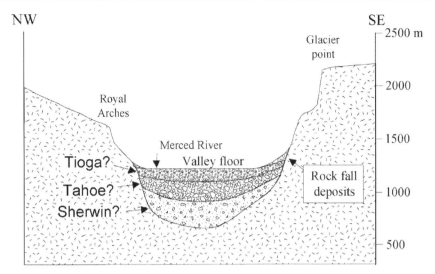

Figure 119. Topographic profile of upper Yosemite Valley, looking towards the northeast. Section shows the geometry of the entire valley, including subsurface strata from the seismic study by Gutenberg and others (1956; conducted in 1935 and 1937). No vertical exaggeration, and the elevations are in meters above sea level. Location of profile line is shown in Fig. 114. Rock fall deposits are probably more common in the subsurface adjacent to the valley walls than shown in the figure.

Extreme topography

Several walls in Yosemite Valley have amazing stretches that tilt well beyond vertical. Of these the most dramatic is Leaning Tower (Fig. 120), located adjacent to Bridal Vein Falls in the west side of the valley. Greater than vertical walls are present on portions of El Capitan. In particular, part of the Salathe Wall to the west of the Nose, and portions of the North American wall exceed vertical. Many short over-vertical bands and overhangs are present at the Royal Arches (Fig. 118). Likewise, the east side of Washington Column has some over vertical terrain. These over-steepened rock formations reflect both the depth of glacial erosion and probably above average rock strength as expressed in the composition of the rock and both the spacing and directions of joint sets.

Another radical form of topography are the spires of Yosemite Valley. The most well known is the Lost Arrow Spire, perched high along the wall to the east of Yosemite Falls (Fig. 121). This spire is about 200 feet high from its point of departure from the main wall, and sits 1,400 feet above the valley floor. Perhaps more frequently overlooked are the remarkable twin spires in the west side of the valley, which are called Higher and Lower Cathedral Spires (Fig. 122). Spires develop in a delicate balancing act where rock of lesser strength, commonly in zones of close-spaced joints, are eroded away leaving behind a tower of stone. Large spires are rare in the regions visited by the John Muir trail, but in other parts of the Sierra Nevada spires have formed. Perhaps the next most dramatic apart from those of Yosemite Valley is the remote, Castle Rock Spire in Kings Canyon National Park, which is approximately 1,000 feet high.

Figure 120. Photograph of Leaning Tower; view towards the southwest. The rock formation is about 1,200-feet high, and the over vertical west face is amongst the steepest walls in the valley. Bridal Veil Falls is at the far-left side of the photograph.

Geology of the John Muir trail 297

Figure 121. Photograph of the Lost Arrow Spire, taken in morning light. A sense of separation is given by comparing the spire to the shadow it casts. The tip of the spire is positioned about 120 feet out from the rest of the wall.

Figure 122. Photograph of Higher and Lower Cathedral Spires as viewed from El Capitan Meadows. The relative heights of the spires may be controlled by a low-angle NE-dipping joint that is largely completely eroded away (left skyline of the Lower Spire). These spires are pock marked with scars where exfoliation and joint-bounded blocks have peeled away.

PLUTONIC ROCKS

Yosemite Valley is in the realm of igneous geology, being made of massive Cretaceous plutons. The eastern side of the valley is eroded into the west side of the Tuolumne Intrusive Suite, and the remainder of the valley has a variety of different aged plutons. The Tuolumne intrusive suite is generally oriented parallel with the Sentinel Granodiorite, which assembles one of the more interesting magmatic bodies in the valley, having complex elongate blocks the El Capitan Granite distributed throughout the NS-elongate pluton. The El Capitan Granite volumetrically comprise a large part of the west side of Yosemite Valley. The El Capitan rock formation gives incredible vertical exposure of large bodies of magmas that underwent molten mixing. This pluton and the Taft Granite were called the Yosemite Intrusive Suite (Bateman, 1992), but it does not have the nested character of suites along the Sierra crest.

El Capitan Magma Mixing

The nearly 3,000 feet of relief exposed at El Capitan secures this formation as one of the largest continuous vertical outcrops in North America, and perhaps one of the preeminent cliffs in the world. It is mainly composed of the El Capitan Granite, intruded at about ~103 Ma (Stern and others, 1981), which is similar in age to the Bullfrog Pluton far south near Rae Lakes. The pluton comprises most of the western half of Yosemite Valley. In terms of dynamic magma processes, the North America Wall of El Capitan contains the most interesting features (Figs. 123 and 124). The heterogeneous nature of El Capitan was noted by Clarence King (1878) who wrote *"A study of this precipice would convince any observer that, whatever may have been the origin of the body as a whole, uniform commingling has failed to take place, and that the sharply defined inclusions are mechanical, not chemical, accidents."* At this time, the formation of granite was largely regarded as the end product of the metamorphic series of schist to gneiss. In other words, the granitic rocks were believed to be magmas that formed because of deep burial. Nonetheless, it is clear from King's description of El Capitan that he noted the hornblende-bearing diorite masses and aplite sills. The dark body of rock that defines the North America shape is a NW-striking dike composed of diorite intruded into the El Capitan Granite such that the two compositions were coexisting magmas (Reid and others, 1983). Textures defining the interactions of two magmas in contact with one another are chilled or quenched margins, pillow structures, and chemical reactions at the contacts. The far right side of the North America Wall is intruded by the younger Taft Granite, which is about 96 Ma (Stern and others, 1981). Calkins (1985) mapped diorite bodies intruded into both the El Capitan and Taft granites, yet these two plutons are about three million years different in age, and the diorite in the North America Wall indicates it was emplaced while the El Capitan Granite was partially molten. This suggests that more than one age of diorite is involved, or what has been correlated to Taft Granite is really older. The latter appears to be the case because Ratajeski and others, (2001) reported five U-Pb dates for the diorites in the Taft and El Capitan Granites that are all between 102 and 105 Ma, which, along with rock textures, indicate the intrusions of El Capitan are mostly coeval. Although, the development of the pluton is significantly more complex (see section below) and largely remains to be studied.

The base of El Capitan is only a ten to fifteen minute hike from the road at El Capitan Meadow (Fig 124) following an unmarked trail. Once at the bottom of the Shield, another ten-minute hike up talus along the base towards the east brings one to the bottom of the North America Wall where some of the intrusive relationships can be examined. One should be aware of where climbing parties are above on the wall and be alert for the infrequent dropped object or rock. The hike to the base is worth it just for seeing a new perspective on the massive El Capitan- the vertical sea of granite. There are many named geographic features on El Capitan that climbers use to navigate by (Fig. 123). Some of these terms are also used in the following section.

A Shattered Monolith

The earlier phase of the El Capitan Granite (Figs. 125A and 130A) made a mass that was foliated and contained widespread small rounded mafic inclusions that may represent mixing within the pluton. This magma became sufficiently rigid to become extensively fractured upon the injection of granodiorite and diorite masses (Figs. 126 and 130B). Despite the monolithic appearance from a distance much of the apparently light colored surfaces are underlain by a mosaic of fragmented granite separated by granodiorite. To some extent these breccia blocks may have accumulated downward. For example, lighter colored blocks are more abundant at the base of the Middle Cathedral Rock across the valley. Many of the injected granodiorite and diorite bodies define wedge shapes, indicating that the El Capitan Granite was extremely hot, if not partially molten (Fig. 127). Some of these wedge-shaped intrusions had the wallrock-dike contracts folded, showing that the material was still hot and able to flow. About 15% of the Shield is composed of granodiorite filling the voids between blocks of shattered granite, difficult to discern from a distance, but prevalent throughout El Capitan once sufficient contrast is allowed for and one understands that the subtle color changes reflects different composition rock (Figs. 128 and 129).

The above described brecciation pattern was overprinted again by brittle style intrusion of diorite that perhaps had a two-phased intrusion process. First, a series of steeply east-dipping en echelon granodiorite to diorite dikes formed an array from top to bottom of the North America Wall (Fig. 130C). These pre-North America dikes were briefly discussed by Ratajeski and others (2001), but they did not consider their extent and en echelon pattern. Second, the North America Wall diorite dike (unit Kid of Calkins, 1985; Fig. 125B) broke through this array, in some cases at least partially isolating large blocks of wall rock within the dike (Figs. 130D and 131). Many of the magma mixing textures between the diorite and the surrounding El Capitan Granite attests to the fact of forceful intrusion during shearing. The partially molten wall rock was capable of moving out of the crystal lattice mush to mix with the invading diorite. In some cases, composite dikes formed with diorite coring the center of aplite (Fig. 132), with the aplite component probably sourced from the wall rock. Some of these dikes record folding. Similar composite sills were described far to the south along the JMT near Bishop Pass (Fig. 72), and yet are relatively uncommon.

Recall that diorite has a higher melting temperature than granite, therefore the difference between the temperature and composition results in appreciable contrast in viscosity. The physical mixing of the two magmas may be inhibited by immiscability, such as when oil and water are stirred together the two fluids remain separated.

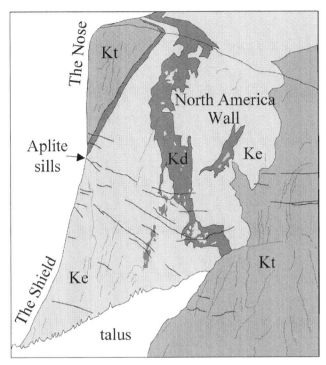

Figure 123. Sketch of the 3,000-foot prow of El Capitan and the North America Wall, which is composed of a diorite dike (Kd) cutting the El Capitan Granite (Ke). The younger Taft Granite (Kt) makes the upper part of the nose of El Capitan, and the far right side of the North American Wall. Heavier lines are aplite sills. Within the diorite dike are numerous blocks of El Capitan Granite that are partially melted. In detail, the diorite was disaggregated into numerous ellipsoid forms throughout the mingling into the more felsic partially molten El Capitan Granite. The diorite that separates the Taft Granite on the top of the Nose from the El Capitan Granite is foliated, and may be an equivalent to the North America Wall diorite dike, but it was highly sheared. This dike can be traced completely around the southwest face of El Capitan. It passes above the Heart, defines part of the Great Roof, intersects the Salathe Wall, and climbs back up to the rim.

Figure 124. Photograph of El Capitan as viewed from the top of Middle Cathedral rock. Trail to base of El Capitan marked with dashed line. EC = El Capitan Meadow, H = the Heart, M = mixed zone of diorite and El Capitan Granite, NA = North America Wall, S = the Shield, N = the Nose, and G = Great Roof. Numbered boxes the show approximate areas of other figures discussing the geology of El Capitan.

Similar diorite dikes also run through the middle of the Three Brothers, and is best exposed up high in the recent rock scars (Fig. 133). These dikes were not mapped by Calkins (1985), and demonstrate, along with many of the above-described textures and compositional variations in El Capitan Granite, that many aspects of Yosemite geology remain to be discovered and studied. Although, one may argue that these dikes are not far different from the overall highly dike-like pattern of intrusion of the highly foliated Sentinel Granodiorite (Fig. 22) into the El Capitan Granite as best exposed in the north-south corridor at Yosemite Falls.

Finally, the El Capitan Granite pluton is about 18 by 60 km in size, thus the exposure within the walls of Yosemite Valley account for but a fraction of the pluton. This being the case, the processes examined so far need not necessarily represent the pluton formation at large.

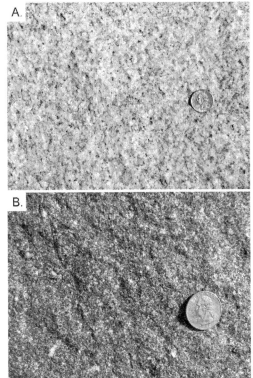

Figure 125. A. Rock photograph of the El Capitan Granite, quarter for scale. **B.** Rock photograph of diorite from the North America Wall; quarter for scale.

Figure 126. Apparent rectangular blocks of El Capitan Granite separated by granodiorite to diorite dikes that have internal mixing textures and evidence for shearing. Image is approximately 5 m high. Darker coloration along the bottom of the photograph is patina or weathered rock. Exposure is along the base of El Capitan along the east flank of the Shield. This is a similar view of the same outcrop that was shown in Figure 6C of Ratajeski and others (2001).

Figure 127. Wedge-shaped dike of granodiorite intruded into lighter El Capitan Granite. Note that both the upper and lower margins of the dike were subsequently folded, providing evidence that the El Capitan Granite was still hot and able to flow. In this case the physical conditions defining the difference between fracturing and flowing was dependent on the strain rate.

Geology of the John Muir trail

Figure 128. Contrast enhanced photograph of part of the Shield of El Capitan. Vertical strips are from water and weathering (location shown in Fig. 124). White aplite dikes diagonal across from upper left to lower right. Masses of lighter gray rock are made of granite (g), whereas regions of gray rock are granodiorite (gd) that invaded the voids between massively shattered granite. Inset box is shown in Fig. 129. View is looking upwards at the Shield, and thus includes foreshortening effects. Height in the image is approximately 250 feet.

Figure 129. Detail of granodiorite filling spaces between shattered blocks of granite (location shown in Fig. 128). Note the apparent isolated block above the prominent aplite dike. Also note how the blocks of granite are separated such that they can be retrofitted without significant rotation.

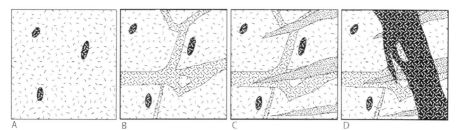

Figure 130. Diagrammatic sketches of the development of part of the El Capitan heterogeneous monolith. **A.** Early granite intrusion was foliated and contained wide-spaced rounded to stretched out mafic inclusions. **B.** Brecciation and intrusion of granodiorite to diorite. **C.** En echelon intrusion of granodiorite to diorite dikes. **D.** Final break through and intrusion of the North America Wall diorite.

Figure 131. Photograph looking at the center of the North America Wall with diorite defining the general shape of the "continent". This is an oblique cut through the diorite dike, and therefore the apparent thickness is exaggerated. The diorite segment making "Baja California" reattaches near the bottom of the photograph, in effect isolating a large piece of the El Capitan Granite. The faint and very light gray relatively thin pre-North America diorite en echelon granodiorite dike array is marked by arrows (these dikes diagonal from lower left to upper right of the photograph).

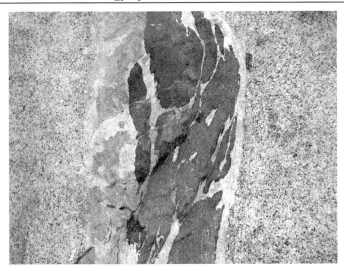

Figure 132. Composite dike of dark diorite and light aplite cross cutting the El Capitan Granite, which is exposed at the base of El Capitan along the east side of the Shield (see Fig. 124 for approximate location). Out of the field of view the dike is folded. Note the quarter along the right side of the dike for scale.

Sills at the Half Dome-Kuna Crest Granodiorite Contact, Bishop's Balcony

The western contact of the Half Dome Granodiorite (~90.5-92.7 Ma, Coleman and others, 2002) had intruded the Kuna Crest Granodiorite (~93 Ma), and is exposed in both the north and south canyon walls of Yosemite. At Glacier Point and Royal Arches, the different composition plutons produce distinct styles of jointing, allowing the contact position to be estimated. In general, the Half Dome Granodiorite is more massive than the Kuna Crest Granodiorite whereas the latter is cut by abundant 45-degree eastward dipping joints. The Half Dome Granodiorite has apparently sent numerous aplitic to pegmatitic sills into the Kuna Crest Granodiorite (Fig. 134). The map by Calkins shows these sills to be from the Half Dome Granodiorite, but they are not composed of granodiorite. From the outcrop geometry, it is tempting to think of the sills as having intruded from right to left. Although, in 3D they may have injected from a direction either in or out of the outcrop face. The sills can be easily examined at the cliff base behind Church Bowl, northeast of Yosemite Village, at a place known to climbers as Bishop's Balcony.

Figure 133. Photograph of the Three Brothers showing the location of a major diorite dike swarm (outlined in thin black lines). The dikes generally lie near the contact between the El Capitan Granite (left side of photograph) and the Sentinel Granodiorite (right side of photograph). The photograph was taken from Taft Point.

Other locations in Yosemite bearing numerous aplite sills are at the North America Wall on El Capitan, Cathedral Rocks, and Upper Yosemite Falls.

Abundant sills along the margins of plutons are a common feature in the Sierra Nevada batholith, and may represent the stress created around a cooling pluton. Others areas with similar well developed sills complexes include the headwaters of Rock Creek near Dade Lake (Sherlock and Hamilton, 1958), at Pine Creek (Bateman, 1965b), and at the cirque south of Silver Pass along the JMT (Lockwood and Lydon, 1975). In most cases, the sills are highly evolved, meaning they were made from the last water and quartz-enriched portion of the magma. Once the pluton is mostly crystallized and begins cooling, it becomes denser and therefore will tend to sink.

This motion is probably responsible for the vertical tension and sill injection at a pluton's margin.

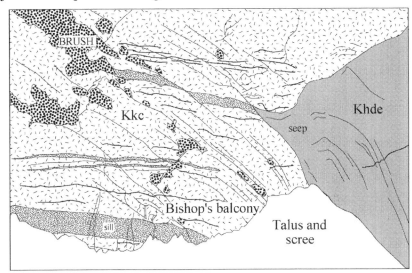

Figure 134. Sketch of the Bishop's Balcony wall, showing the contact area between the Kuna Crest (Kkc) and Half Dome (Khde) Granodiorites. Outcrop is at left side of Royal Arches, north side of Yosemite Valley, as seen from near Church Bowl. The Kuna Crest Granodiorite has a sill complex that probably originated from the adjacent Half Dome Granodiorite. The sills (marked by heavier lines and dark hatch pattern) are aplite to pegmatite, and the thicker sills are layered. The sills are cut by 45-degree NE-dipping joint set. These joints are not as abundant in the Half Dome Granodiorite. The sills formed under tensile stress that was vertically oriented, and the inclined joint set indicates a major change from this earlier stress field.

SUMMARY ON PART OF THE LATE CRETACEOUS MAGMATISM

The various Sierra Nevada intrusive units described in this guide were injected both coeval and sequentially throughout the final phase of batholith consolidation. Several of the plutons interacted with the continent margin parallel right-lateral faults, such as the Sierra Crest shear zone, so that their map form reflects this deformation. In an attempt to show the temporal development, Figure 135 illustrates a series of time frames for the main intrusive suites crossed by the John Muir Trail.

The sequence begins with the earlier plutons of the Tuolumne and John Muir intrusive suites at about 93 to 90 Ma. Immediately west of the John Muir Intrusive Suite the circa 90 Ma Mount Givens Granodiorite was also emplaced. Bateman (1992) included this pluton in the John Muir Intrusive Suite, however, it does not share the obliquity of elongation and therefore perhaps formed by a different mechanism. In the 88 to 89 Ma period, the bulk of the John Muir Intrusive Suite (Fig. 65) was emplaced following an en echelon or stacked geometry, a form certainly related to right-lateral shearing during plutonism. Right-lateral strike-slip deformation in the Cretaceous continental margin is in accord with known oblique tectonic plate convergence directions determined by paleo-magnetitic studies (Engebretson and others, 1985). In contrast, both the Tuolumne and Mount Whitney Intrusive Suites formed as concentric multiple-pulse magma chambers where younger intrusions were centered at the core of the older injection (Bateman, 1992). Most of the Tuolumne Intrusive Suite was emplaced throughout the 93 to 88 Ma interval (Coleman and others, 2004), before the initial opening of the Mount Whitney Intrusive Suite. Both of these latter suites have right-lateral faults that with additional studies may prove more important for their role in pluton emplacement. However, multiple and prolonged dike injection during formation of the intrusive suites is supported by the recent geochronology and limited described field localities (Glazner and others, 2002; Coleman and others, 2004), but has not come to terms with an integrated explanation with the bulk chemical and petrological zonation of the nested plutons. From about 85 to 83 Ma magmatic activity was more abundant to the south. Note that this series of time frames focuses on the three dominant intrusive suites along the John Muir Trail. Additional activity in the batholith was happening both to the north and south, and for the earlier time frames, to the west. At the end of time frame 81 to 83 Ma older granitic rocks and major roof pendants occupy the white regions.

The duration of the Sierra Crest shear zone, a ductile fault that runs the length of the batholith, may span the development of all the intrusive suites (Greene and Schweickert, 1995; Tikoff and Saint Blanquat, 1997; Tikoff and others, 1999), and other subparallel large displacement faults were probably also active (Lahren and Schweickert, 1990). More importantly, and perhaps in the spirit of older studies that used the concept of intrusive epochs (e.g., Evernden and Kistler, 1970), examining the big picture development of the

Cretaceous batholith hints at a deeper level (lower-crust to asthenosphere depth) processes of mass magma generation during the subduction of oceanic crust. Separation of the major intrusive suites into discrete entities or units is almost periodic, and is suggestive of piercement structures generated by the rise of diapirs from a deeper magma reservoir. The spaced character of the plutons certainly reflects the emplacement mechanism, and probably not the extent nor duration of the underlying source melt that probably formed more of a continuous ribbon-like body beneath much of the western continental margin of the Americas. This is especially apparent when examining the distribution of Cretaceous batholiths, of which the Sierra Nevada batholith is but a small segment of the larger system. Spaced-out intrusive suites and plutons along strike of the batholith is also the case for the distribution and timing of intrusive activity in the Coastal batholith of Peru. In fact, major volcanoes in modern arcs are also spaced apart, such as the volcanic islands in the Aleutian arc, or stratovolcanos of the Cascade arc.

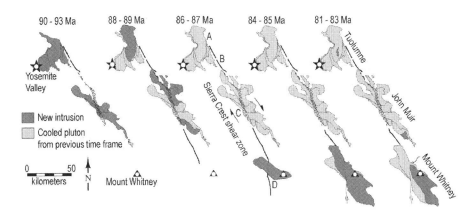

Figure 135. Generalized emplacement timing of plutons into intrusive suites. A. Cascade Lake shear zone, B. Gem Lake shear zone, C. Rosy-Finch shear zone, and D. Kern Canyon Fault.

APPENDIX 1: AGE DATA

Plutons crossed by the John Muir Trail	Age	Reference
Johnson Granite Porphyry	81 Ma, 84 Ma, 85.4	6, 5, 8
Cathedral Peak Granodiorite	86 Ma, 88.1	6, 8
Half Dome Granodiorite	90.6 to 92.6 Ma	8, 11
Intruded by the Cathedral Peak gd		
Kuna Crest Granodiorite	91 Ma, ~93 Ma	6, 11
Mono Creek Granite	88 Ma	1
Round Valley Granodiorite	89 Ma	6
Cut by the Mono Creek Granite		
Graveyard Peak Luecogranite	99 Ma	6
Intruded by Lake Edison Granodiorite		
Lake Edison Granodiorite	90 Ma	7
Cut by the Round Valley Granodiorite		
Lamarck Granodiorite	89.6 Ma, 91.9 Ma	6, 9
Intruded by Lake Edison and Mount Givens Granodiorites		
Evolution Basin Granite	<89.6 Ma - 85 Ma	4
Intrudes the Lamarck Granodiorite		
Sheared granodiorite of Goddard Pendant	>158 Ma	
Cartridge Pass Granodiorite	81 Ma	4
Contains inclusions of Lamarck Granodiorite		
McDoogle Pluton	95 Ma	10
Cut by the Lamarck Granodiorite		
Twin Lakes Pluton	<165 Ma, >148 Ma	
Cut by IDS, intrudes the Tinemaha Granodiorite		
Independence Dike Swarm (IDS)	148 Ma	3
Tinemaha Granodiorite	165 Ma	2
Intruded by IDS		
White Fork Pluton	156 Ma	2
Cut by IDS		
Bullfrog Pluton	103 Ma	2
Diamond Pluton	>103 Ma, <156 Ma	
Cut by mafic dikes (IDS?)		
Paradise Granodiorite	83-86 Ma	2
Whitney Granodiorite	83 Ma	2

(1) Bateman (1992)
(2) Chen and Moore (1982)
(3) Chen and Moore (1979)
(4) Evernden and Kistler (1970)
(5) Fleck and others (1996)
(6) Stern and others (1981)
(7) Tobisch and others (1995)
(8) Coleman and Glazner (1997)
(9) Coleman and others (2000)
(10) Mahan and Bartley (2000)
(11) Coleman and others (2002)

APPENDIX 2: TRAIL ACCESS POINTS

There are numerous ways to gain access to the John Muir Trail and to hike smaller portions of the trail. In general, eastern side access points have short and steep approaches whereas those along the western side are lengthy and gradual (Table 3). For the main trailheads below, I have made a brief list of starting elevations, gain to the major pass, and distance to the JMT accompanied with GPS waypoints. Various possible combinations of entry and exit points can be developed, particularly if the hiking party has two vehicles. Those planning on hitching a ride and using the trailheads belonging to the more difficult east side approaches, known as the gruesome four, should be prepared to hike the additional mileage to the highway.

TABLE 3 COMPARISON OF JMT APPROACHES

Trailhead	Elev.	Pass/approach	Elev.	Gain	Miles	Ft./mi.
Whitney Portal	8,361'	Trail Crest	13,680'	5,239'	8.2	638
Whitney Portal	8,361'	Whitney summit	14,494'	6,134'	11	558
*Shepherd Pass trailhead	6,431'	Shepherd Pass	12,008'	5,577'	16	349
Roads End-Cedar Grove	5,035'	Bubbs Creek	9,515'	4,480'	14	320
Onion Valley	9,200'	Kearsarge Pass	11,823'	2,623'	9	291
*Baxter Pass trailhead	6,037'	Baxter Pass	12,304'	6,267'	13	482
Roads End-Cedar Grove	5,035'	Woods Creek	8,466'	3,430'	13.4	256
*Black Rock Spg. Rd.	4,600'	Sawmill Pass	11,347'	6,747'	13.5	500
*Taboose Creek	5,455'	Taboose Pass	11,380'	5,925'	10	592
South Lake	9,755'	Bishop Pass	11,972'	2,217'	13.5	158
Rock Creek	10,400'	Mono Pass	12,050'	1,650'	17	97
North Lake	9,360'	Piute Pass	11,423'	2,063'	18	115
Florence Lake	7,343'	Evolution Valley	7,860'	517'	11	47
Bear Creek dam	7,650'	Bear Creek	8,910'	1,610'	6.75	239
Lake Thomas Edison	7,650'	Mono Creek	7,900'	250'	6	47
Cold Water	9,122'	Duck Lake Pass	10,790'	1,668'	7	238
Devils Postpile	~7,600'	NA	0	0	0	0
Agnew Meadows	8,300'	Shadow Lake	8,800'	500'	4	125
Cathedral Lakes trailhead	8,590'	NA	0	0	0	0
Happy Isles	4,035'	Cathedral Pass	9,700'	5,665'	15.7	287

* gruesome 4

Whitney Portal to Mount Whitney. East side access, Map 1. The Whitney Portal trailhead (waypoint 389167E, 4049569N) is the southern starting point to gain access to the John Muir Trail, which technically begins at the summit of Mount Whitney (4,416.8 m, 14,491.9'). A paved road extends 20.9 km (13 miles) from Lone Pine to the trailhead, and the trail starts at 2,550 m (8,365') elevation. From Whitney Portal, the ascent to Trail Crest Pass (13,600') and then onto Mount Whitney is 17.7 km (11 miles) accompanied with a gain of 1,870 m (6,134'). At the trailhead there is a small store, potable water, but no telephone. Most backpackers carrying the heavy load for the trek of the entire John Muir trail will need to spend the night at the upper trail camp before ascending to the summit.

Shepherd Pass. East side access, Map 4. Almost hate to include this probable torture approach, but for completeness, the trailhead is at 385670E, 4064925N at the top of a four-wheel drive road, approximately at the 1,960 m (6,431') elevation. Shepherd Pass, located at 379886E, 4059158N, lies at about 3,660 m (12,008') elevation, accounting for a total gain of about 1,700 m (5,577'). The trail intersects with the JMT near Tyndall Creek, at 376008E, 4055534N. It is about 16 miles from the trailhead to the JMT.

Bubbs Creek (Cedar Grove). West side access, Map 7. The Bubbs Creek access joins the JMT to the north of Forester Pass and south of Glen Pass (waypoint 374065E, 4068994N). The trail begins near Cedar Grove (1,535 m, 5,035') in Kings Canyon at the end of Highway 180, called Road's End (waypoint 358856E, 4073108N). After 22.5 km (14 miles) the trail joins the JMT to the south of Glen Pass, near Vidette Meadow, which lies to the north of Forester Pass.

Onion Valley -Kearsarge Pass. East side access, Map 7. The Onion Valley trailhead (2,804 m, 9,200'; waypoint 380501E, 4070320N) has paved access running 24.1 km (15 miles) westward from Independence. The trail goes over Kearsarge Pass (3,603 m, 11,823'; waypoint 377227E, 4070413N) and joins the JMT after total of 14.5 km (9 miles), entering to the south of Glen Pass and to the north of Forester Pass. This is a popular trailhead, having relatively easy approach to the JMT (gain of 2,623'), and may require additional effort to secure a wilderness permit.

Baxter Pass. East side access, Map 9. The Baxter Pass route has the greatest elevation gain off all the possible approaches to the JMT. Its trailhead is located at 384377E, 4078187N, elevation 1,840 m (6,037'), climbs to the pass at elevation of ~3,750 m (12,304'; waypoint 377382E, 4077284N), and joins the JMT at 374586E, 4077296N just north of Dollar Lake. The hike is about 13-miles long.

Woods Creek (Cedar Grove). West side access, Map 9. Woods Creek trail begins at the same trailhead at Road's End as described above for Bubbs Creek, but follows the canyon to the north, passing Mist Falls, and ascending along the north side of the Middle Fork of the Kings River and Woods Creek. The trail meets the JMT (waypoint 371817E, 4081632N) to the south of Pinchot Pass and north of Glen Pass, having a total elevation gain of 1,045 m (3,430') over 21.6 km (13.4 miles). Grants Grove has a store, visitor center, and showers.

Sawmill Pass. East side access, Map 10. The trail to Sawmill Pass (3,153 m, 10,346'; waypoint 378503E, 4082569N) joins the JMT to the south of Pinchot Pass (waypoint 375323E, 4084768N), and is approached from a trailhead reached from Black Rock Springs Road that is 19.3 km (12 miles) north of Independence. The trailhead elevation is at 1,402 m (4,600'; waypoint 385243E, 4088666N). This is one of the more difficult approaches to the JMT.

Taboose Pass. East side access, Map 11. The trailhead is located along Taboose Creek Road, located 22.5 km (14 miles) to the north of Independence and 19.3 km (12 miles) south of Big Pine. The trail starts, at 1,663 m (5,455'; waypoint 381994E, 4096545N), and climbs 11.3 km (7 miles) to Taboose Pass (3,469 m, 11,380'; waypoint 374236E, 4108541N), then meets the JMT in about 16 km (10 miles) total distance from the trailhead. This JMT trail junction is north of Pinchot Pass and south of Mather pass. The trail makes a split joining the JMT at two points; the southern entry point is at 372005E, 4091441N. This is a difficult approach, and for an exit point leaves one in the desert with little chance of transportation if arrangements have not been made for in advance.

Bishop Pass (South Lake). East side access, Map 16. The Bishop Pass trail starts at South Lake, elevation 2,973 m (9,755'; waypoint 361082E, 4114589N), located at the end of a paved road that runs 35.4 km (22 miles) from Bishop. The trailhead to the John Muir Trail is 22.5 km (14 miles) long, passing Bishop Pass (3,649 m, 11,972'; waypoint 362820E, 4108541N), which has an elevation difference of 675.7 m (2,217') and is one of the access points with the least amount of gain along the east side. If this is used as an exit point, the climb up from the JMT to Dusy Basin and Bishop Pass is grueling. Entry point with JMT at 358395E, 4106305N.

Piute Pass (North Lake). East side access, Map 22. Piute Pass route comes from a trailhead located at North Lake (2,853 m, 9,360'; waypoint 355714E, 4121127N), which has a paved road from Bishop. Piute Pass is at 3,482 m (11,423'; waypoint 350709E, 4122502N), and the total distance from the trailhead to the JMT is 30 km (18 miles. JMT entry point at 337427E, 4121211N.

Florence Lake. West side access, Map 22. The Florence Lake trailhead (~2,240 m, ~7,350'), sharing the access road 168 with Lake Thomas Edison, provides a reasonable approach into Evolution Valley (entry at JMT waypoint 334934E, 4121350N). From the JMT to Florence Lake is 17.6 km (11 miles). About 6.5 km (4 miles) of hiking may be trimmed off by taking the small ferryboat to the far side of the lake ($8 one-way or $15 round-trip) that runs every half-hour. See www.muirrailranch.com/FL-Ferry.html.

Lake Thomas Edison (Bear and Mono Creeks). West side access. Lake Thomas Edison (2,329.5 m, 7,643') lies at the end of a long, narrow, twisty route 168 that climbs up out of Fresno. Both the Mono and Bear Creek segments of the JMT may be approached from the lake.

From the trail junction at Mono Creek (Map 26, waypoint 329743E, 4142191N), it is about 9.7 km (6 miles) to the Vermilion Valley resort. A ferry service crosses the lake to the Vermilion Valley Resort, charging a similar fee as for Florence Lake.

The Bear Creek trailhead is reached from a four-wheel drive service road south of Lake Thomas Edison that leads to a diversion dam (waypoint 325068E, 4133771N). This 10.1 km (6.25 miles) long trail joins the JMT in Map 25 at waypoint 332909E, 4137190N.

Mono Pass (Rock Creek). East side access, Map 27. The Rock Creek trailhead is at the road's end at Mosquito Flat (3,169 m, 10,400'; waypoint 345510E, 4144379N), which is approached from a paved road leading west from Tom's Place along Highway 395. The trailhead to the JMT covers 27.4 km (17 miles), going over Mono Pass at an elevation of 3,673 m (12,050'; waypoint 343237E, 4143350N), and providing for a relatively low elevation gain for an entry point to the JMT. JMT entry point at 331031E, 4143236N.

Mammoth Lakes. East side access. Several jumping off points lie around the town of Mammoth Lakes. The most direct are the trailheads in the Devils Postpile National Monument, either at the Devils Postpile parking lot for north bound hikers (intersection with JMT at 315955E, 4164844N), or at Reds Meadow (~2,316 m, ~7,600') for a south bound direction (Map 32, trailhead waypoint 316804E, 4164844N). For the north section of the trail, another trailhead is located at Agnew Meadows (2,530 m, 8,300'). This trail descends into the valley, and then climbs up to JMT at Shadow Lake (2,682 m, 8,800'), covering a distance of 5.6 km (3.5 miles). All these trailheads may be reached by car driving west of Mammoth Lakes and over Minaret summit, however, vehicle access is restricted to before 7 am and after 8 pm. During the day a shuttle service leaving at the Mammoth Ski resort is mandatory. The Devils Postpile National Monument now charges a $5 dollars user fee per person, which also covers use of the shuttle bus. The regulations, fees, etc. change from year to year, but the information is generally available online.

To the south of Mammoth Lakes, a convenient trailhead south of Mary Lake at the Cold Water parking area (waypoint 324400E, 4162157N), bypasses the limited access and fees for the Devils Postpile region, and shaves off a long ascent to the south of Reds Meadow. The trail over Duck Lake Pass (3,289 m, 10,790'; waypoint 326726E, 4158472N) joins the JMT (waypoint 326099E, 4156208N) to the south of Duck Lake (Map 29), covering about 11.3 km (7 miles).

Tuolumne Meadows and Cathedral Lakes trailheads. West or East access. Located along Highway 120 in Yosemite National Park, the Tuolumne Meadows and Cathedral Lakes trailheads, and various other locations in the Tuolumne Meadows area, provides convenient trail access at a general elevation of 2,618 m (8,590'). In the Tuolumne Meadows area is a small store, post office, a greasy spoon grill, gas station, and telephones, which may or may not be a welcomed relief to tired northbound hikers. The Tuolumne Meadows trailhead is east of the store and southeast of Lembert Dome, with a trailhead parking lot at the ranger permit kiosk (Map 39, waypoint 293738E, 4194535N). The trail intersects the JMT at 294789E, 4193588N. The Cathedral Lakes trailhead lies west of the Visitor Center and has limited parking along Highway 120 (Map 40), and intersects the JMT at 290444E, 4194120N.

Happy Isles, Yosemite Valley. West or East access, Map 44. Happy Isles is the northern end of the John Muir Trail (waypoint 274633E, 4178845N), located at the eastern end of Yosemite Valley at an elevation of 1,230 m (4,035'). Yosemite Valley is approached from the east by Highway 120 leaving Highway 395 at Lee Vining, or from the west by Highways 120, 140, or 41. From the Happy Isles trailhead the JMT immediately climbs up to Cathedral Pass (2,956 m, 9,700'), gaining 1,720.5 m (5,645') over a distance of 25.3 km (15.7 miles).

REFERENCES

Abbott, R.E., and Louie, J.N., 2000, Depth to bedrock using gravimetry in the Reno and Carson City, Nevada, area basins: Geophysics, v. 65, p. 340-350.

Ague, J.J., and Brimhall, G.H., 1988, Magmatic arc symmetry and distribution of anomalous plutonic belts in the batholiths of California: Effects of assimilation, crustal thickness, and depth of crystallization: Geological Society of America Bulletin, v. 100, p. 912-927.

Alonso Olazabala, A., Carracedo, M., and Aranguren, A., 1999, Petrology, magnetic fabric and emplacement in a strike-slip regime of a zoned peraluminous granite: the Campanario-La Haba pluton, Spain: *in* Castro A., Fernández C. & Vigneresse J.L., eds., Geological Society Special Publication, v. 168, p. 177-190.

Anderson, E.M., 1951, The dynamics of faulting and dyke formation with applications to Britain: 2nd ed. Edinburgh, Oliver and Boyd, 206 p.

Atwater, T., 1970, Implications of plate tectonics of the Cenozoic tectonic evolution of western North America: Geological Society of America Bulletin, v. 81, p. 3513-3535.

Axelrod, D.I., 1957, Late Tertiary floras and the Sierra Nevada uplift: Geological Society of America Bulletin, v. 68, p. 19-45.

Axelrod, D.I., 1962, Post-Pliocene uplift of the Sierra Nevada of California: Geological Society of America Bulletin, v. 73, p. 183-198.

Axen, G.J., Taylor, W.J., and Bartley, J.M., 1993, Space-time patterns and tectonic controls of Tertiary extension and magmatism in the Great Basin of the western United States: Geological Society of America Bulletin, v. 105, p. 56-76.

Bachman, S.B., 1978, Pliocene-Pleistocene break-up of the Sierra Nevada-White-Inyo Mountains block and formation of Owens Valley: Geology, v. 6, p. 461-463.

Bahat, D., Grossenbacher, K., and Karasaki, K., 1999, Mechanism of exfoliation joint formation in granitic rocks, Yosemite National Park: Journal of Structural Geology, v. 21, p. 85-96.

Bailey, R.A., 1989, Geologic Map of the Long Valley Caldera, Mono-Inyo Craters Volcanic Chain, and vicinity, eastern California: U.S. Geol. Survey Miscellaneous Investigation Series, I-1933, scale 1:62,500.

Bateman, P.C., 1992, Plutonism in the Central Sierra Nevada, California: U.S. Geological Survey Professional Paper, 1483, 186 p.

Bateman, P.C., 1965a, Geologic Map of the Black Cap Mountain Quadrangle, Fresno County, California, CQ-428.

Bateman, P.C., 1965b, Geology and Tungsten mineralization of the Bishop district, California, with a section on gravity study of Owens Valley by Pakiser, L.C. and Kane, M.F., and a section on seismic profile by Pakiser, L.C.: U.S. Geological Survey Professional Paper 470, 208 p.

Bateman, P.C., and Chappell, B.W., 1979, Crystallization, fractionation, and solidification of the Tuolumne Intrusive Series, Yosemite National Park, California: Geological Society of America Bulletin, v. 90, p. 465-482.

Bateman, P,C., Kistler, R.W., Peck, D.L., and Busacca, A., 1983, Geologic Map of the Tuolumne Meadows Quadrangle, CQ-1570.

Bateman, P.C., and Wahrhaftig, C., 1966, Geology of the Sierra Nevada: Bulletin of the California Division of Mines and Geology, v. 190, p. 107-172.

Bateman, P.C., and Moore, J.G., 1965, Geologic Map of Mount Goddard Quadrangle, Fresno and Inyo Counties, California, CQ-429.

Beard, J.S., and Day, H.W., 1987, The Smartville intrusive complex, Sierra Nevada, California: the core of a rifted volcanic arc: Geological Society America Bulletin, v. 99, p. 779-791.

Becker, G.F., 1891, Structure of a portion of the Sierra Nevada: Geological Society of America, v. 2, p. 49-74.

Benson, L.V., May, H.M., Antweiler, R.C., Brinton, T.I., Kashgarian, M., Smoot, J.P., and Lund, S.P., 1998, Continuous lake-sediment records of glaciation in the Sierra Nevada between 52,600 and 12,500 ^{14}C yr B.P.: Quaternary Research, v. 50, p. 113-127.

Bergbauer, S., and Martel, S.J., 1999, Formation of joints in cooling plutons: Journal of Structural Geology, v. 21, p. 821-835.

Bergeron, L.K., 1992, Structural evolution of the Mesozoic section of the Mount Morrison roof pendant: unpublished MS Thesis, University of California, Santa Cruz, 107 p.

Berry, M.E., 1997, Geomorphic analysis of late Quaternary faulting on Hilton Creek, Round Valley and Coyote warp faults, east-central Sierra Nevada, California, USA: Geomorphology, v. 20, p. 177-195.

Best, M.G., and Christiansen, E.H., 1991, Limited extension during peak Tertiary volcanism, Great Basin of Nevada and Utah: Journal of Geophysical Research, v. 96, p. 13,509-13,528.

Bierman, P., Gillespie, A., Whipple, K., and Clark, D., 1991, Quaternary geomorphology and geochronology of Owens Valley, California: Geological Society of America Field Trip Guide, *in* Walawender, M.J., and Hanan, B.B., Geological excursions in southern California and Mexico, Department of Geological Sciences, San Diego State University, p. 199-223.

Birkeland, P.W., Burke, R.M., and Walker, A.L., 1980, Soils and subsurface rock-weathering features of Sherwin and pre-Sherwin glacial deposits, eastern Sierra Nevada, California: Geological Society of America Bulletin, v. 91, p. 238-244.

Birman, J.H., 1964, Glacial geology across the crest of the Sierra Nevada, California: Geological Society of America Special Paper 75, 79 p.

Bischoff, J.L., Menking, K.M., Fitts, J.P., and Fitzpatrick, J.A., 1997, Climatic oscillations 10,000-155,000 yrs B.P. at Owens Lake, California, reflected in glacial rock flour abundances and lake salinity in core OL-92: Quaternary Research, v. 48, p. 313-325.

Blackwelder, E., 1931, Pleistocene glaciation in the Sierra Nevada and Basin Ranges; Geological Society of America Bulletin: v. 42, no. 4, p. 865-922.

Blandy, T.D., and Sparks, R.S.J., 1992, Petrogenesis of mafic inclusions in granitoids of the Adamello Massif, Italy: Journal of Petrology, v. 33, p. 1039-1104.

Bogen, N.L., 1985, Stratigraphic and sedimentological evidence of a submarine island-arc volcano in the lower Mesozoic Peñon Blanco and Jasper Point Formations, Mariposa County, California: Geological Society of America Bulletin, v. 96, p. 1322-1331.

Bowen, N.L., 1928, The Evolution of the Igneous Rocks: Princeton, N.J., Princeton University Press- reissued by Dover Publications, 333 p.

Bradbury, J.P., 1997, A diatom record of climate and hydrology for the past 200 ka from Owens Lake, California with comparison to other Great Basin records: Quaternary Science Reviews, v. 16, p. 203-219.

Brocklehurst, S.H., and Whipple, K.X., 2002, Glacial erosion and relief production in the eastern Sierra Nevada, California: Geomorphology, v. 42, p. 1-24.

Burgmann, R., and Pollard, D.D., 1994, Strain accommodation about strike-slip fault discontinuities in granitic rock under brittle-to-ductile conditions: Journal of Structural Geology, v. 16, p. 1655-1674.

Burke, R.M., and Birkeland, P.W., 1979, Reevaluation of multiparameter relative dating techniques and their application to the glacial sequence along the eastern escarpment of the Sierra Nevada, California: Quaternary Research, v. 11, p. 21-51.

Burkins, D.L., Blum, J.D., Brown, Kevin, Reynolds, R.C., and Erel, Y., 1999, Chemistry and mineralogy of a granitic, glacial soil chronosequence, Sierra Nevada Mountains, California: Chemical Geology, v. 162, p. 1-14.

Bursik, M., and Sieh, K., 1989, Range front faulting and volcanism in the Mono Basin, eastern California: Journal of Geophysical Research, v. 94, p. 15,587-15,609.

Busby-Spera, C.J., and Saleeby, J.R., 1990, Intra-arc strike slip fault exposed at batholithic levels in the southern Sierra Nevada, California: Geology, v. 18, p. 255-259.

Busby-Spera, C.J., 1988, Speculative tectonic model for the early Mesozoic arc of the southwest Cordilleran United States: Geology, v. 16, p. 1121-1125.

Calkins, C., 1985, Bedrock Geologic Map of Yosemite Valley, Yosemite National Park, California, U.S. Geological Survey Miscellaneous Investigation Series, I-1639.

Carl, B.S., Glazner, A.F., and Bartley, J.M., 1997, Composite Independence dikes, eastern California: Geological Society of America Abstracts with Programs, v. 29, p. A-391.

Catchings, R.D., 1992, A relation among geology, tectonics, and velocity structure, western to central Nevada Basin and Range: Geological Society of America Bulletin, v. 104, p. 1178-1192.

Chamberlain, C.P., and Poage, M.A., 2000, Reconstructing the plaeotopography of mountain belts from the isotopic composition of authigenic minerals: Geology, v. 28, p. 115-118.

Chase, C.G., and Wallace, T.C., 1986, Uplift of the Sierra Nevada of California: Geology, v. 14, p. 730-733.

Chen, J.H., 1977, Zircon geochronology of the Sierra Nevada Batholith: unpublished Ph.D. dissertation, University of California, Santa Barbara, 210 p.

Chen, J.H. and Moore, J.G., 1982, Uranium-Lead isotopic ages from the Sierra Nevada Batholith, California: Journal of Geophysical Research, v. 87, p. 4761-4784.

Chen, J.H., and Moore, J.G., 1979, Late Jurassic Independence dike swarm in eastern California: Geology, v. 7, p. 129-133.

Chen, J.H., and Tilton, G.R., 1976, Isotopic lead investigations on the Allende carbonaceous chrondrite: Geochimica et Cosmochimica Acta, v. 40, p. 635-643.

Christensen, M.N., 1966, Late Cenozoic crustal movement in the Sierra Nevada of California: Geological Society of America Bulletin, v. 77, p. 163-182.

Christensen, M.N., Gilbert, C.M., Lajoie, K.R., and Al-Rawi, Y., 1969, Geological-geophysical interpretation of Mono Basin, California-Nevada: Journal of Geophysical Research, v. 74, p. 5221-5239.

Clark, D.H., and Gillespie, A.R., 1997, Timing and significance of late-glacial and Holocene cirque glaciation in the Sierra Nevada, California: Quaternary International, v. 38/39, p. 21-38.

Cobbing, J., 2000, The Geology and Mapping of Granite Batholiths: Springer, Berlin, 141 p.

Coleman, D.S., Gray, W., and Glazner, A.F., 2004, Rethinking the emplacement and evolution of zoned plutons: geochronologic evidence for incremental assembly of the Tuolumne Intrusive Suite, California: Geology, v. 32, p. 433-436.

Coleman, D.S., Gray, W., and Glazner, A.F., 2002, U-Pb geochronologic evidence for incremental filling of the Tuolumne intrusive suite magma chamber: Geological Society of America Abstracts with Programs, v. 34, p. 269.

Coleman, D.S., Carl, B.S., Glazner, A.F., and Bartley, J.M., 2000, Cretaceous dikes within the Jurassic Independence dike swarm in eastern California: Geological Society of America Bulletin, v. 112, p. 504-511.

Coleman, D.S., and Glazner, A.F., 1997, The Sierra Crest magmatic event: rapid formation of juvenile crust during the Late Cretaceous in California: International Geology Review, v. 39, p. 768-787.

Coleman, D.S., Glazner, A.F., Miller, J.S., Bradford, K.J., Frost, T.P., Joye, J.L., and Bachl, C.A., 1995, Exposure of a Late Cretaceous layered mafic-felsic magma system in the central Sierra Nevada batholith, California: Contributions to Mineralogy and Petrology, v. 120, p. 129-136.

Collins, W.J., Richards, S.R., Healy, B.E., and Ellison, P.I., 2000, Origin of heterogenous mafic enclaves by two-stage hybridisation in magma conduits (dykes) below and in granitic magma chambers: Transactions of the Royal Society of Edinburgh: Earth Sciences, v. 91, p. 27-45.

Conrad, J.E., 1993, Late Cenozoic tectonics of the southern Inyo Mountains, eastern California: unpublished Masters thesis, San Jose State University, 84 p.

Cox, B.F., 1982, Stratigraphy, sedimentology, and structure of the Goler Formation (Paleocene), El Paso Mountains, California; implications for Paleogene tectonism on the Garlock fault zone: Ph.D. dissertation, University of California, Riverside, 296 p.

Cruden, A.R., 1988, Deformation around a rising diapir modeled by creeping flow past a sphere: Tectonics, v. 7, p. 1091-1101.

Dalrymple, G.B., 1964, Potassium-argon dates of three Pleistocene interglacial basalt flows from the Sierra Nevada, California: Geological Society of America Bulletin, v. 75, p. 753-758.

Dalrymple, G.B., 1963, Potassium-argon dates of some Cenozoic volcanic rocks of the Sierra Nevada, California: Geological Society America Bulletin, v. 74, p. 379-390.

Davis, M., Teyssier, C., and Tikoff, B., 1995, Dextral shearing in the Cascade Lake shear zone, Tuolumne Intrusive Suite, Sierra Nevada, California: Geological Society of America Abstracts with Programs, v. 27, p. 222.

Day, H.W., Moores, E.M., and Tuminas, A.C., 1985, Structure and tectonics of the northern Sierra Nevada: Geological Society America Bulletin, v. 96, p. 436-450.

Didier, J., 1973, Granites and their enclaves: the bearing of enclaves on the origin of granites: Elsevier Scientific Publishing Company, Amsterdam, 393 p.

Dixon, J.M., 1975, Finite strain and progressive deformation in models of diapiric structures: Tectonophysics, v. 28, p. 89-124.

Dixon, T.H., Miller, M., Farina, F., Wong, H., and Johnson, D., 2000, Present-day motion of the Sierra Nevada block and some tectonic implications for the Basin and Range province, North America Cordilleran: Tectonics, v. 19, p. 1-24.

Dodge, F.C.W., and Moore, J.G., 1968, Occurrence and composition of biotite from the Cartridge Pass Pluton of the Sierra Nevada batholith, California: U.S. Geological Survey Professional Paper 600-B, p. B6-B10.

Dorn, R.I., Turrin, B.D., Jull, A.J.T., Linick, T.W., and Donahue, D.J., 1987, Radiocarbon and Cation-Ratio ages for rock varnish on Tioga and Tahoe morainal boulders of Pine Creek, eastern Sierra Nevada, California, and their paleoclimatic implications: Quaternary Research, v. 28, p. 38-49.

Duffield, W.A., Bacon, C.R., and Dalrymple, G.B., 1980, Late Cenozoic volcanism, geochronology, and structure of the Coso Range, Inyo County, California: Journal of Geophysical Research, v. 85, p. 2381-2404.

Dumitru, T.A., 1990, Subnormal Cenozoic geothermal gradients in the extinct Sierra Nevada magmatic arc: consequences of Laramide and post-Laramide shallow-angle subduction: Journal of Geophysical Research, v. 95, p. 4925-4941.

Dunne, G.C., and Walker, J.D., 1993, Age of Jurassic volcanism and tectonism, southern Owens Valley region, east-central California: Geological Society of America Bulletin, v. 105, p. 1223-1230.

Dupras, D.L., 1985, Life beneath the Temblor Sea; Sharktooth Hill, Kern County, California: California Geology, v. 38, p. 147-154.

Edelman, S.H., and Sharp, W.D., 1989, Terranes, early faults, and pre-Late Jurassic amalgamation of the western Sierra Nevada metamorphic belt, California: Geological Society America Bulletin, v. 101, p. 1420-1433.

Engebretson, D.C., Cox, A., and Gordon, R.G., 1985, Relative motions between oceanic and continental plates in the Pacific basin: Geological Society of America Special Paper 206, 59 p.

Evernden. J.F., and Kistler, R.W., 1970, Chronology of emplacement of Mesozoic batholithic complexes in California and Western Nevada: U.S. Geological Survey Professional Paper 623, 43 p.

Fairbanks, H.W., 1898, The great Sierra Nevada fault scarp: Popular Science, v. 52, p. 609-621.

Fiske, R.S., and Tobisch, O.T., 1994, Middle Cretaceous ash-flow tuff and caldera-collapse deposit in the Minarets Caldera, east-central Sierra Nevada, California: Geological Society of America Bulletin, v. 106, p. 582-593.

Fleck, R.J., Kistler, R.W., and Wooden, J.L., 1996, Geochronological complexities related to multiple emplacement history of the Tuolumne Intrusive Suite, Yosemite National Park, California: Geological Society of America Abstracts with Program, v. 28, p. 65-66.

Flinn, J.E., and Reid, J.B., 1986, Coupled Magma Mixing and Crystal Sorting in the Johnson Granite Porphyry, Yosemite National Park, California: EOS abstract, v. 67, no. 44, p. 1269.

Frost, T.P., and Mahood, G.A., 1987, Field, chemical, and physical constraints on mafic-felsic magma interaction in the Lamarck Granodiorite, Sierra Nevada, California; Geological Society of America Bulletin, v. 99, p. 272-291.

Gardner, J.V., Mayer, L.A., and Hughs Clark, J.E., 2000, Morphology and processes in Lake Tahoe (California-Nevada): Geological Society of America Bulletin, v. 112, p. 736-746.

Gilbert, G.K., 1877, Report on the geology of the Henry Mountains: Publication of the Powell Survey, United States Department of Interior, Washington DC., 160 p.

Gillespie, A.R., 1982, Quaternary glaciation and tectonism in the southeastern Sierra Nevada, Inyo County, California: Ph.D. dissertation, California Institute of Technology, Pasadena, CA, 695 p.

Glazner, A.F., Taylor, R.Z., Bartley, J.M., and Gray, W., 2002, Dike assembly of the Tuolumne intrusive suite, Yosemite National Park, California: Geological Society of America Abstracts with Programs, v. 34, p. 269.

Glazner, A.F., Bartley, J.M., and Carl, B.S., 1999, Oblique opening and noncoaxial emplacement of the Jurassic Independence dike swarm, California: Journal of Structural Geology, v. 21, p. 1275-1283.

Glazner, A.F., and Bartley, J.M., 1984, Timing and tectonic setting of Tertiary low-angle normal faulting and associated magmatism in the southwestern United States: Tectonics, v. 3, p. 385-396.

Greene, D.C., 1995, Stratigraphy, structure, and tectonic significance of the northern Ritter Range pendant, eastern Sierra Nevada, California [Ph.D. dissertation.]: University of Nevada, Reno, 270 p.

Greene, D.C., Stevens, C.H., and Wise, J.M., 1997, The Laurel-Convict fault, eastern Sierra Nevada, California: A Permo-Triassic left-lateral fault, not a Cretaceous intrabatholithic break: Geological Society of America Bulletin, v. 109, p. 483-488.

Greene, D.C., and Schweickert, R.A., 1995, The Gem Lake shear zone: Cretaceous dextral transpression in the northern Ritter Range pendant, eastern Sierra Nevada, California: Tectonics, v. 14, p. 945-961.

Gutenberg, B., Buwalda, J.P., and Sharp, R.P., 1956, Seismic explorations on the floor of Yosemite Valley, California: Geological Society of America Bulletin, v. 67, p. 1051-1078.

Guyton, B., 1998, Glaciers of California: California Natural History Guides, 59, University of California Press, Berkeley, 197 p.

Hack, J.T., 1960, Interpretation of erosional topography in humid temperate regions: American Journal of Science, v. 258, p. 80-97.

Hannah, J.L., and Moores, E.M., 1986, Age relationships and depositional environments of Paleozoic strata, northern Sierra Nevada, California: Geological Society America Bulletin, v. 97, p. 787-797.

Hanson, R.E., 1991, Quenching and hydroclastic disruption of andesitic to rhyolitic intrusions in a submarine island-arc sequence, northern Sierra Nevada, California: Geological Society America Bulletin, v. 103, p. 804-816.

Hanson, R.E., Saleeby, J.B., and Schweickert, R.A., 1988, Composite Devonian island-arc batholith in the northern Sierra Nevada, California: Geological Society America Bulletin, v. 100, p. 446-457.

Haq, B.U., Hardenbol, J., and Vail, P.R., 1987, Chronology of fluctuating sea-levels since the Triassic: Science, v. 235, p. 1156-1166.

Harbor, J.M., 1992, Numerical modeling of the development of U-shaped valleys by glacial erosion: Geological Society of America Bulletin, v. 104, p. 1364-1375.

Henry, C.D., and Perkins, M.E., 2001, Sierra Nevada-Basin and Range transition near Reno, Nevada: two-stage development at 12 Ma and 3 Ma: Geology, v. 29, p. 719-722.

Herzig, C.T., and Sharp, W.D., 1992, The Sullivan Creek terrane: A composite Jurassic arc assemblage, western Sierra Nevada metamorphic belt, California: Geological Society America Bulletin, v. 104, p. 1292-1300.

Hirt, W.H., 2007, Petrology of the Mount Whitney intrusive suite, eastern Sierra Nevada, California: implications for the emplacement and differentiation of composite felsic intrusions: Geological Society of America Bulletin, v. 119, p. 1185-1200.

Hirt, W.H., 1989, The petrological and mineralogical zonation of the Mount Whitney Intrusive Suite, eastern Sierra Nevada, California: unpublished Dissertation, Santa Barbara, University of California, 278 p.

House, M.A., Wernicke, B.P., Farley, K.A., Dumitru, T.A., 1997, Cenozoic thermal evolution of the central Sierra Nevada, California, from (U-Th)/He thermochronometry: Journal of Geophysical Research, v. 102, p. 11,745-11,763.

Huber, K.N., 1989, The Geologic Story of Yosemite National Park: Yosemite Association, Yosemite National Park, 63 p.

Huber, K.N., 1981a, Amount and timing of Late Cenozoic uplift and tilt of the central Sierra Nevada: U.S. Geological Survey Professional Paper 1197, 28 p.

Huber, K.N., 1981b, Late Cenozoic evolution of the Tuolumne River, central Sierra Nevada, California: Geological Society of America Bulletin, v. 102, p. 102-115.

Huber, K.N., Bateman, P.C., and Wahrhaftig, C., 1989, Geologic Map of Yosemite National Park and vicinity, California: U.S. Geological Survey Miscellaneous Investigation Series, I-1874.

Huber, K.N., and Eckhardt, W.W., 1985, Devils Postpile Story: The Sequoia National History Association, Three Rivers, California, 30 p.

Huber, K.N., and Rinehart, D.C., 1967, Cenozoic Volcanic Rocks of the Devils Postpile Quadrangle, Eastern Sierra Nevada California: U.S. Geological Survey Professional Paper 554-D, 21 p.

Huber, N.K., and Rinehart, C.D., 1965, Geologic Map of the Devils Postpile Quadrangle, Sierra Nevada, California, U.S. Geological Survey, CQ-437.

Jackson, M.P.A., and Talbot, C.J., 1986, External shapes, strain rates, and dynamics of salt structures: Geological Society of America Bulletin, v. 97, p. 305-323.

Johnson, S.E., Albertz, M., and Paterson, S.R., 2001, Growth rates of dike-fed plutons; are they compatible with observations in the middle and upper crust? Geology, v. 29, p. 727-730.

King, Clarence, 1871, Mountaineering in the Sierra Nevada: republished 1935 by W.W. Norton and Company, Inc., New York, 320 p.

King, Clarence, 1878, Systematic Geology, Volume 1, United States Geological Exploration of the Fortieth Parallel, Washington, 803 p.

Kistler, R.W., 1966, Geologic Map of the Mono Craters Quadrangle, U.S. Geological Survey Geologic Map, CQ-464.

Kistler, R.W., and Swanson, S.E., 1981, Petrology and geochronology of metamorphosed volcanic rocks and a middle Cretaceous volcanic neck in the east-central Sierra Nevada, California: Journal of Geophysical Research, v. 86, p. 10,489-10,501.

Kistler, R.W., and Peterman, Z.E., 1973, Variations in Sr, Rb, K, Na, and initial Sr^{87}/Sr^{86} in Mesozoic granitic rocks and intruded wallrocks in central California: Geological Society of America Bulletin, v. 84, p. 3489-3512.

Knopf, A., 1918, A geologic reconnaissance of the Inyo Range and the eastern slope of the southern Sierra Nevada, California: U.S. Geological Survey Professional Paper 110, 130 p.

Lahren, M.M, and Schweickert, R.A., 1990, Evidence of uppermost Proterozoic to Lower Cambrian Miogeoclinal rocks and the Mojave-Snow Lake Fault: Snow Lake pendant, central Sierra Nevada, California: Tectonics, v. 9, p. 1585-1608.

Lawson, A.C., 1904, The geomorphology of the upper Kern basin: University of California, Department of Geological Sciences Bulletin, v. 3, p. 291-376.

Le Conte, J., 1886, A post-Tertiary elevation of the Sierra Nevada shown by the river beds: American Journal of Science, 3rd ser., v. 32, p. 167-181.

Le Conte, Joseph, 1873, On some of the ancient glaciers of the Sierra: American Journal of Science and Arts, 3rd series, vol. 5, p. 325-342.

Lee, J., Spencer, J., and Owen, L., 2001, Holocene slip rates along the Owens Valley fault, California; implications for the recent evolution of the Eastern California shear zone: Geology, v. 29, p. 819-822.

Lindgren, W., 1911, The Tertiary gravels of the Sierra Nevada of California: U.S. Geological Survey Professional Paper 73, 226 p.

Lockwood, J.P., and Moore, J.G., 1979, Regional deformation of the Sierra Nevada, California, on conjugate micro fault sets: Journal of Geophysical Research, v. 84, p. 6041-6049.

Lockwood, J.P., and Lydon, P.A., 1975, Geological Map of the Mount Abbot Quadrangle: U.S. Geological Survey, CQ-1155.

Lubetkin, L.K.C., and Clark, M.M., 1988, Late Quaternary activity along the Lone Pine fault, eastern California: Geological Society of America Bulletin, v. 100, p. 755-766.

Lueddecke, S.B., Pinter, N., and Gans, P., 1998, Plio-Pleistocene ash falls, sedimentation, and range-front faulting along the White-Inyo mountain front, California: Journal of Geology, v. 106, p. 511-522.

Mahan, K.H., and Bartley, J.M., 2000, Emplacement of the McDoogle pluton, central Sierra Nevada, via sheeted diking into a contractional shear zone: Geological Society of America abstracts with programs, v. 32, p. A26.

Manhes, G., Göpel, C., and Allegre, C.J., 1986, Lead isotopes in Allende inclusions: the oldest known solar material: Terra Cognita, v. 6, p. 173.

Marsh, B.D., 1982, On the mechanics of igneous diapirism, stoping, and zone melting: American Journal of Science, v. 282, p. 808-855.

Martel, S.J., Pollard, D.D., and Segall, P., 1988, Development of simple strike-slip fault zones, Mount Abbot quadrangle, Sierra Nevada, California: Geological Society of America Bulletin, v. 100, p. 1451-1465.

Martinson, D.G., Pisias, N.G., Hays, J.D., Imbrie, J., Moore, T.C., Jr., and Shackleton, N.J., 1987, Age dating and the orbital theory of the ice ages: Development of a high resolution 0 to 360,000-year chromstratigraphy: Quaternary Research, v. 26, p. 1-29.

Matthes, F.E., 1965, Glacial Reconnaissance of Sequoia National Park, California: U.S. Geological Survey Professional Paper 504-A, p. 58.

Matthes, F.E., 1960, Reconnaissance of the Geomorphology and Glacial Geology of the San Joaquin Basin, Sierra Nevada, California: U.S. Geological Survey Professional Paper 329, p. 62.

Matthes, F.E., 1937, The geologic history of Mt. Whitney: Sierra Club Bulletin, v. 22, p. 1-18.

Matthes, F.E., 1930, Geologic History of the Yosemite Valley: U.S. Geological Survey Professional Paper 160, 137 p.

Mayo, E.B., 1941, Deformation in the interval Mount Lyell-Mount Whitney, California: Geological Society America Bulletin, v. 52, p. 1001-1084.

McNulty, B.A., Tobisch, O.T., Cruden, A.R., and Gilder, S., 2000, Multistage emplacement of the Mount Givens pluton, central Sierra Nevada batholith, California: Geological Society of America Bulletin, v. 122, p. 119-135.

McNulty, B.A., Farber, D.L., Wallace, G.S., Lopez, R., and Palacios, O., 1998, Role of plate kinematics and plate-slip-vectors partitioning in continental magmatic arcs; evidence from the Cordillera Blanca, Peru: Geology, v. 26, p. 827-830.

McNulty, B.A., Tong, W., and Tobisch, O.T., 1996, Assembly of a dike-fed magma chamber, the Jackass Lakes pluton, central Sierra Nevada, California: Geological Society of America Bulletin, v. 108, p. 926-940.

Menking, K.M., Bischoff, J.L., Fitzpatrick, J.A., Burdette, J.W., and Rye, R.O., 1997, Climatic/hydrologic oscillations since 155,000 yr B.P. at Owens Lake, California, reflected in abundance and stable isotope composition of sediment carbonate: Quaternary Research, v. 48, p. 58-68.

Menzies, J., 2002, Modern and Past Glacial Environments: Butterworth Heinemann, Oxford, 543 p.

Moore, J.G., 2000, Exploring the highest Sierra: Stanford University Press, Stanford, California, 427 p.

Moore, J.G., 1981, Geologic map of the Mount Whitney Quadrangle: U.S. Geological Survey, CQ-1545.

Moore, J.G., 1963, Geology of the Mount Pinchot Quadrangle, southern Sierra Nevada, California; U.S. Geological Survey Bulletin 1130, p. 152.

Moore, J.G., and Dodge, F.C.W., 1980, Late Cenozoic volcanic rocks of the southern Sierra Nevada, California: I. Geology and petrology: Summary: Geological Society of America Bulletin, v. 91, p. 515-518.

Moore, J.G., and duBray, E., 1978, Mapped offset on the right-lateral Kern Canyon Fault, Southern Sierra Nevada, California: Geology, v. 6, p. 205-208.

Moore, J.G., and Hopson, C.A., 1961, The Independence Dike Swarm in eastern California: American Journal of Science, v. 259, p. 241-259.

Muir, John, 1874, Studies in the Sierras: Overland Monthly- republished by the Sierra Club, San Francisco, p. 103.

Muir, John, 1873, Discovery of glaciers in the Sierra Nevada: American Journal of Science and Arts, 3rd series, vol. 5, p. 69-71.

Muir, John, 1869, My First Summer in the Sierra: Boston and New York, Houghton Mifflin company, Riverside Press, Cambrage, 354 p.

Pachell, M.A., and Evans, J.P., 2002, Growth, linkage, and termination processes of a 10-km-long strike-slip fault in jointed granite: the Gemini fault zone, Sierra Nevada, California: Journal of Structural Geology, v. 24, p. 1903-1924.

Pachell, M.A., Evans, J.P., and Taylor, W.L., 2003, Kilometer-scale kinking of crystalline rocks in a transpressive convergent setting, central Sierra Nevada: Geological Society of America Bulletin, v. 115, p. 817-831.

Pakiser, L.C., 1976, Seismic exploration of Mono Basin, California: Journal of Geophysical Research, v. 81, p. 3607-3618.

Pakiser, L.C., Kane, M.F., and Jackson, W.H., 1964, Structural geology and volcanism of Owens Valley region, California- A geophysical study: U.S. Geological Survey Professional Paper 438, 68 p.

Paterson, S.R., and Vernon, R.H., 1995, Bursting the bubble of ballooning plutons: A return to nested diapirs emplaced by multiple processes: Geological Society of America Bulletin, v. 107, p. 1356-1380.

Paterson, S.R., Vernon, R.H., and Tobisch, O.T., 1989, A review of criteria for the identification of magmatic and tectonic foliations in granitoids: Journal of Structural Geology, v. 11, p. 349-363.

Peck, D.L., 2002, Geologic map of the Yosemite quadrangle, central Sierra Nevada, California: U.S. Geological Survey, Geological Investigations Series Map I-2751, scale 1:62,500.

Penck, A., and Brückner, E., 1909, Die Alpen im Eiszeitalther: Tauchnitz, Leipzig, Bd., v. 1-3, 1199 p.

Petford, N., and Atherton, M.P., 1992, Granitoid emplacement and deformation along a major crustal lineament: the Cordillera Blanca, Peru: Tectonophysics, v. 205, p. 171-185.

Piccoli, P.M., Candela, P.A., Frank, M.R., and Jugo, P.J., 1997, Anatomy, textural development and mafic-silicate chemistry of late-stage granite dikes, Tuolumne Intrusive Suite, CA: Geological Society of America Abstracts with Programs, v. 29, p. A-390.

Phillips.F.M., Zreda, M.G., Smith, S.S., Elmore, D., Kubik, P.W., and Sharma, P., 1990, Cosmogenic chlorine-36 chronology for glacial deposits at Bloody Canyon, eastern Sierra Nevada: Science, v. 248, p. 1529-1532.

Pitcher, W.S., 1997, The nature and origin of granite: 2nd edition, Chapman and Hall, Glasgow, 387 p.

Pitcher, W.S., 1991, Synplutonic dykes and mafic enclaves, *in* Didier, J., and Barbarin, B., eds., Enclaves and granite petrology: Developments in Petrology, v. 13, p. 383-391.

Poage, M.A., and Chamberlain, C.P., 2002, Stable isotope evidence for a pre-Middle Miocene rain shadow in the western Basin and Range: Implications for the paleotopography of the Sierra Nevada: Tectonics, v. 21, p. 16.1-16.10.

Pollard, D.D., and Aydin, A., 1988, Progress in understanding jointing over the past century: Geological Society of America Bulletin, v. 100, p. 1181-1204.

Putnam, W.C., 1962, Late Cenozoic geology of McGee Mountain, Mono County, California: University California Publications Geological Science, v. 40, p. 181-218.

Ramberg, H., 1972, Theoretical models of density stratification and diapirism in the Earth's crust: Journal of Geophysical Research, v. 77, p. 877-889.

Ramelli, A.R., Bell, J.W., dePalo, C.M., and Yount, J.C., 1999, Large-magnitude, late Holocene earthquake on the Genoa fault; west-central Nevada and eastern California: Bulletin of the Seismological Society of America, v. 89, p. 1458-1472.

Ratajeski, K., Glazner, A.F., and Miller, B.V., 2001, Geology and geochemistry of mafic to felsic plutonic rocks in the Cretaceous intrusive suite of Yosemite Valley, California: Geological Society of America Bulletin, v. 113, p. 1486-1502.

Reid, J.B., Murray, D.P., Hermes, O.D., and Steig, E.J., 1993, Fractional crystallization in granites of the Sierra Nevada: How important is it?: Geology, v. 21, p. 587-590.

Reid, J.B.Jr., Evans, O.C., and Fates, D.G., 1983, Magma mixing in granitic rocks of the central Sierra Nevada, California: Earth and Planetary Science Letters, v. 66, p. 243-261.

Reid, S.A., and Cox, B.F., 1989, Early Eocene uplift of southernmost San Joaquin Basin, California: American Association of Petroleum Geologists Bulletin, vol.73, no.4, abstract, pp.549-550, Apr 1989.

Rinehart, D.C., and Ross, D.C., 1964, Geology and Mineral Deposits of the Mount Morrison Quadrangle, Sierra Nevada, California: U.S. Geological Survey Professional Paper 385, p. 106.

Russell, I.C., 1889, Quaternary history of Mono County, California: eighth annual report, U.S. Geological Survey, pt. 1, p. 302-304.

Russell, I.C., 1887, Notes on the faults of the Great Basin of the eastern base of the Sierra Nevada: Bulletin of the Philosophical Society, Washington, v. 9, p. 5-7.

Sack, D., 1992, New wine in old bottles: the historiography of a paradigm change: Geomorphology, v. 5, p. 251-263.

Sams, D.B., and Saleeby, J.B., 1988, Geology and petrotectonic significance of crystalline rocks of the southernmost Sierra Nevada, California: *in* Ernst, W.G., Metamorphism and crustal evolution of the Western United States, Rubey Volume, v. 7, p. 865-893.

Saleeby, J.B., Kistler, R.W., Longiaru, S, Moore, J.G., and Nokleberg, W.J., 1990, Middle Cretaceous metavolcanic rocks in the Kings Canyon area, central Sierra Nevada, California: Geological Society of America Memoir 174, p. 251-270.

Sarna-Wojcicki, A.M., Pringle, M.S., and Wijbrans, J., 2000, New $^{40}Ar/^{39}Ar$ age of the Bishop Tuff from multiple sites and sediment rate calibration for the Matuyama-Brunhes boundary: Journal of Geophysical Research, v. 105, p. 21,431-21,443.

Schaffer, J.P., 1997, The Geomorphic Evolution of the Yosemite Valley and Sierra Nevada Landscapes, solving the riddles in the rocks: Wilderness Press, Berkeley, California, 388 p.

Schmeling, H., Cruden, A.R., and Marquart, G., 1988, Finite deformation in and around a fluid sphere moving through a viscous medium: implications for diapiric ascent: Tectonophysics, v. 149, p. 17-34.

Schweickert, R.A., and Lahren, M.M., 1993, Tectonics of the east-central Sierra Nevada- Saddlebag Lake and northern Ritter Range pendants: *in* Lahren, M.M., Trexler, J.H., and Spinosa, C., eds., Crustal Evolution of the Great Basin and

Sierra Nevada: Geological Society of America, University of Nevada, Reno, Guidebook, p. 313-351.

Schopf, J.W., 1993, Micro fossils of the early Archean Apex Chert; new evidence of the antiquity of life: Science, v. 260, p. 640-646.

Segall, P., McKee, E.H., Martel, S.J., and Turrin, B.D., 1990, Late Cretaceous age of fractures in the Sierra Nevada batholith: Geology, v. 18, p. 1248-1251.

Shackleton, N.J., and Opdyke, N.D., 1976, Oxygen-isotope and paleomagnetic stratigraphy of the Pacific core V28-239, late Pliocene to latest Pleistocene: Geological Society of America Memoir 145, p. 449-464.

Sharp, W.D., Tobisch, O.T., and Renne, P.R., 2000, Development of Cretaceous transpressional cleavage synchronous with batholith emplacement, central Sierra Nevada, California: Geological Society of America Bulletin, v. 122, p. 1059-1066.

Sharp, R.P., 1972, Pleistocene glaciation, Bridgeport basin, California: Geological Society of America Bulletin, v. 83, p. 2233-2260.

Sharp, R.P., 1969, Semiquantitative differentiation of glacial moraines near Convict Lake, Sierra Nevada, California: Journal of Geology, v. 77, p. 68-91.

Sharp, R.P., 1968, Sherwin Till-Bishop Tuff Geological Relationships, Sierra Nevada, California: Geological Society of America Bulletin, v. 79, p. 351-364.

Sharp, R.P., and Birman, J.H., 1963, Additions to classical sequence of Pleistocene glaciations, Sierra Nevada, California: Geological Society of America Bulletin, v. 74, p. 1079-1086.

Sherlock, D.G., and Hamilton, W., 1958, Geology of the north half of the Mt. Abbot quadrangle, Sierra Nevada, California: Geological Society of America Bulletin, v. 69, p. 1245-1268.

Sisson, T.W., 1991, Field geochemical, and experimental studies of aluminous arc magmas: unpublished Dissertation, Massachusetts Institute of Technology, Cambridge, 267 p.

Sisson, T.W., Grove, T.L., and Coleman, D.S., 1996, Hornblende gabbro sill complex at Onion Valley, California and a mixing origin for the Sierra Nevada batholith: Contributions to Mineralogy and Petrology, v. 126, p. 81-108.

Small, E.E., and Anderson, R.S., 1995, Geomorphically driven Late Cenozoic rock uplift in the Sierra Nevada, California: Science, v. 270, p. 277-280.

Starr, W.A., 1984, Starr's guide to the John Muir Trail and the High Sierra region: 12th edition, Sierra Club Books, 224 p.

Stern, T.W., Bateman, P.C., Morgan, B.A., Newall, M.F., and Peck, D.L., 1981, Isotopic U-Pb Ages of Zircon from Granitoids of the Central Sierra Nevada, California: U.S. Geological Survey Professional Paper 1185, 17 p.

Stevens, C.H., and Greene, D.C., 1999, Stratigraphy, depositional history, and tectonic evolution of the Paleozoic continental-margin rocks in roof pendants of the eastern Sierra Nevada, California: Geological Society of America Bulletin, v. 111, p. 919-933.

Stock, G.M., and Anderson, R.S., 2002, Testing late Cenozoic uplift of the Sierra Nevada, California, using cave-derived river incision rates: Geological Society of America Abstracts with Programs, v. 34, p. 161.

Strahler, A.N., 1950, Equilibrium theory of erosional slopes approached by frequency distribution analysis: American Journal of Science, v. 248, p. 673-696.

Streckeisen, A.L., 1976, To each plutonic rock its proper name: Earth Science Review, v. 12, p. 1-34.

Taylor, S.R., 1975, Lunar Science: a Post Apollo View: Oxford, Pergamon Press, 372 p.

Tikoff, B.T., Saint Blanquat, M., and Teyssier, C., 1999, Translation and the resolution of the pluton space problem: Journal of Structural Geology, v. 21, p. 1109-1117.

Tikoff, B.T., and de Saint Blanquat, M., 1997, Transpressional shearing and strike-slip partitioning in the Late Cretaceous Sierra Nevada magmatic arc, California: Tectonics, v. 16, p. 442-459.

Tikoff, B.T., and Teyssier, C., 1992, Crustal-scale en echelon "P-shear" tensional bridge: A possible solution to the batholithic room problem: Geology, v. 20, p. 927-930.

Titus, S.J., Clark, R., and Tikoff, B., 2002, Preservation of a volcanic conduit: Johnson Granite Porphyry, Sierra Nevada, California: Geological Society of America Abstracts with Program, v. 34, p. 374.

Tobisch. O.T., Fiske, R.S., Saleeby, J.B., Holt, E., and Sorensen, S.S., 2000, Steep tilting of metavolcanic rocks by multiple mechanisms, central Sierra Nevada, California: Geological Society of America Bulletin, v. 122, p. 1043-1058.

Tobisch, O.T., Saleeby, J.B., Renne, P.R., McNulty, B.A., and Tong, W., 1995, Variations in deformation fields during development of a large volume magmatic arc, central Sierra Nevada, California: Geological Society of America Bulletin, v. 107, p. 148-166.

Tobisch, O.T., Renne, P.R., and Saleeby, J.B., 1993, Deformation resulting from regional extension during pluton ascent and emplacement, central Sierra Nevada, California: Journal of Structural Geology, v. 15, p. 609-628.

Tobisch, O.T., Paterson, S.R., Saleeby, J.B., and Geary, E.E., 1989, Nature and timing of deformation in the Foothills terrane, central Sierra Nevada, California: its bearing on orogenesis: Geological Society of America Bulletin, v. 101, p. 401-413.

Tobisch, O.T., Saleeby, J.B., and Fiske, R.S., 1986, Structural history of continental volcanic arc rocks, eastern Sierra Nevada, California; a case for extensional tectonics: Tectonics, v. 5, p. 65-94.

Tobisch, O.T., Fiske, R.S., Sacks, S., Taniguchi, D., 1977, Strain in metamorphosed volcaniclastic rocks and its bearing on the evolution of orogenic belts: Geological Society of America Bulletin, v.88, p. 23-40.

Unruh, J.R., 1991, The uplift of the Sierra Nevada and implications for late Cenozoic epeirogeny in the western Cordillera: Geological Society of America Bulletin, v. 103, p. 1395-1404.

Vignaud, P., and 21 others, 2002, Geology and palaeontology of the upper Miocene Toros-Menalla hominid locality, Chad: Nature, v. 418, p. 152-155.

Wakabayashi, J., and Sawyer, T.L., 2001, Stream incision, tectonics, uplift, and evolution of topography of the Sierra Nevada, California: Journal of Geology, v. 109, p. 539-562.

Webb, R.W., 1936, Kern Canyon fault, southern Sierra Nevada: Journal of Geology, v. 44, p. 631-638.

Weinberg, R.F., and Podladchikov, Y., 1994, Diapiric ascent of magmas through power-law crust and mantle: Journal of Geophysical Research, v. 99, p. 9543-9560.

Wernicke, B., Clayton, R., Ducea, M., Jones, C.H., Park, S., Ruppert, S., Saleeby, J.S., Snow, J.K., Squires, L., Fliedner, M., Jiracek, G., Keller, R., Klemperer, S., Luetgert, J., Malin, P., Miller, K., Mooney, W., Oliver, H., and Phinney, R., 1996, Origin of high mountains in the continents: The southern Sierra Nevada: Science, v. 271, p. 190-193.

Whitney, J.D., 1865, Geological Survey of California, Geology vol. 1, Legislature of California, 498 p.

Wieczorek, G.F., Snyder, J.B., Waitt, R.B., Morrissey, M.M., Uhrhammer, R.A., Harp, E.L., Norris, R.D., Bursik, M.I., and Finewood, L.G., 2000, Unusual July 10, 1996, rock fall at Happy Isles, Yosemite National Park, California: Geological Society of America Bulletin, v. 112, p. 75-85.

Wilde, S.A., Valley, J.W., Peck, W.H., and Graham, C.M., 2001, Evidence from detrital zircons for the existence of continental crust and oceans on the Earth 4.4 Gyr ago: Nature, v. 409, p. 175-178.

Wilson, J., and Grocott, J., 1999, The emplacement of the granitic Las Tazas complex, northern Chile: the relationship between local and regional strain: Journal of Structural Geology, v. 21, p. 1513-1523.

Winnett, T., and Morey, K., 1984, Guide to the John Muir Trail: Wilderness Press, Berkeley, 104 p.

Wise, J.M., 1996, Structure and Stratigraphy of the Convict Lake block, Mount Morrison pendant, eastern California: [Master's thesis] Reno, University of Nevada, 321 p.

Wolfe, J. A., Schorn, H. E., Forest, C. E., and Molnar, P., 1997, Paleobotanical Evidence for High Altitudes in Nevada During the Miocene: Science, v. 276, p. 1672-1675.

Wood, D.J., and Saleeby, J.B., 1997, Late Cretaceous-Paleocene extensional collapse and disaggregation of the southernmost Sierra Nevada batholith: International Geology Review, v. 39, p. 973-1009.

Zoback, M.L., McKee, E.H., Blackely, R.J., and Thompson, G.A., 1994, The northern Nevada rift: regional tectono-magmatic relations and middle Miocene stress direction: Geological Society of America Bulletin, v. 106, p. 371-382.

INDEX
(page; map number)

Andesite of Mammoth Pass 224, 226; 31-32
Aplite 50, 51-52, 110
Arrowhead Lake 137; 8
Aspen Meadow 187; 21

Banner Peak 199, 244
Basalt of the Buttresses 228, 229; 32
Basalt of Devils Postpile 230-231; 32
Baxter Pass 139, 314
Bear Creek 191-193, 195, 316; 24-26
Bear Dome Quartz Monzonite 189-190; 23-24
Bear Ridge 181, 195, 199; 26
Bighorn Plateau 113, 115, 119, 195; 3
Big Pete Meadow 172; 16
Bishop's Balcony 307, 309
Bishop Pass 172, 315
Bishop Tuff 224, 226; 31-32
Black Divide 172, 174
Boreal Plateau 101
Bubbs Creek 119, 127-131, 314; 6
Budd Creek 266; 40
Bullfrog Lake 131-132; 7
Bullfrog Pluton 126-**130**, 140; 5-9

Caltech Peak 119; 5
Cartridge Pass Pluton 155, 156, 158; 11-13
Cataract Creek 164; 14
Cathedral Lakes 271, 317; 41
Cathedral Pass 271; 41
Cathedral Peak 270; 40
Cathedral Peak Granodiorite 253; 38-42
Cedar Grove 17, 130, 140, 314
Center Peak 128
Chagoopa Plateau 102-103, 105
Charlotte Lake 131-132; 7
Chatter marks 76, 79
Chief Lake 211; 28
Cirque 76
Cleavage 22

Colby Meadow 182
Columbia Finger 271; 41
Consulation Lake ; 1
Crabtree Meadow ; 2
Crater Meadow 220; 31
Crater Mountain 68, 149

Dana Plateau 102
Davis Lakes ; 35
Deadman Pass Volcanics 226; 34
Deer Creek 219; 30
Deer Meadow 164-165; 14
Devils Postpile National Mon. 225-227, 236; 32
Diamond Mesa 102, 114, 119; 5
Diamond Pluton 137
Diapir 58
Disappointment Peak 163
Dollar Lake 137, 139
Donohue Pass 251
Dragon Pluton 130, 136, 137
Duck Lake 217, 316
Dusy Branch 172

El Capitan 298, 288-291, 301-302
Emerald Lake 243; 35
Equigranular 29
Evolution Basin Granite **162**, 164-165, 173; 12-19
Evolution Creek 182-183; 19-20
Evolution Lake 181-182; 19
Evolution Meadow 183; 20
Evolution Valley 182-183, 315; 20

Fairview Dome 45, 247, 270; 40
Faults 41
Fin Dome 130, 133, 137; 8
Fish Creek 212; 28-30
Fissures 288, 290
Florence Lake 188, 315
Folds 45
Foliation 49
Forester Pass 119; 5
Frost wedging 43, 111

Garnet Lake 241; 34
Glacial steps 276
Glacier Point 279, 280, 307; 44

Gladys Lake 237; 34
Glen Pass 133; 7
Goddard pendant 162, 172, **185**; 17, 21-23
Golden Bear dike 126-127; 5
Golden Staircase 164; 14
Grizzly Peak ; 44
Grouse Meadow 172; 15
Guitar Lake 110; 1

Half Dome 273, 276
Half Dome Granodiorite 252; 38-40, 41-44
Happy Isles 279-280; 44
Hardness 21
Heart Lake 189, 191; 23
Helen Lake 173, 176; 17
Higher Catherdral Spire 297
Hitchcock Lakes ; 1

Illilouette Falls ; 44
Inconsolable Granodiorite 153, 163; 13
Independence Dike Swarm 143
Inyo Mountains 18, 72, 105, 143
Island Pass 249; 35
Italy Pass 193

John Muir Intrusive Suite 94, **148**, 310
Johnston Meadow 236; 33
Johnston Granite Porphyry 266, **276**; 39-40
Joints 42
Junction Peak 119; 5

Kearsarge Pass 132, 314
Kearsarge Pinnacles 130; 6
Kern Canyon fault 127, 311
Kern Plateau 99
Kern River 72, 102-103, 105-106
Kings River
      Middle Fork 103, 133, 158, 162, 165, 167, 172, 178; 15-16
      South Fork 149, 153, 154, 158; 11-12
Kings Canyon 96, 168, 277

Kuna Crest Granodiorite 245, **247**; 34-38

Lake McDermand 181; 18
Lake Marjorie 149, 153-154; 11
Lake Thomas Edison 195, 316
Lake Edison Granodiorite 192, **193**; 24-26
Lamarck Granodiorite 136, 145-**147**, 153-155, 157, 158, 163, 164-167, 181-184; 10-25
Langille Peak 172; 16
Leaning Tower 295-296
LeConte Canyon 171, 172, 176, 277; 16-17
Lembert Dome 80, 247, 254, 270
Liberty Cap 276, 278; 44
Little Pete Meadow 172; 16
Little Yosemite Valley 265, 275, 278; 44
Long Meadow 271, 272; 41
Long Valley Caldera 66, 69, 213, 220, 223, 224, 226, 229, 230, 237
Lost Arrow Spire 295, 297
Lower Cathedral Spire 296, 297
Lower Vidette Meadow 131; 7
Lyell Canyon 251, 254; 37-39

Mafic inclusions 47
Mammoth Mountain 220-222, 226, 229, 231, 251
Marie Lake 191, 192; 24
Mather Pass 158, 162; 13
McClure Meadow 182; 20
McDoogle Pluton **145**, 146, 148, 149, 153, 156; 10-11
McGee stage glaciation 82
Merced River 275-280, 293; 44
Middle Palisade peak 163; 13
Minaret Creek 236; 32
Mono Basin stage glaciation 83
Mono Creek 196, 198, 199; 26
Mono Creek Granite **197**-199, 212-214, 216-217, 219, 220, 222, 224, 228-230, 236; 25-33
Mono Pass 196, 316
Moraine Dome 272; 43
Mount Cedric Wright 144, 149; 10

Mount Darwin 101
Mount Hitchcock 110
Mount Huxley 181; 18
Mount Morrison pendant 66, 186, 191, 199, 211-**213**, 215, 218, 219, 237, 241; 28-30
Mount Spencer 182; 18
Mount Warlow 181; 18
Mount Whitney 8, 9, 14, 16, 67, 68, 71, 99-103, 105, 106; 1
Mount Whitney Intrusive Suite 94, 99, **106**, 108, 109, 112, 126, 193, 310
Muir Pass 173, **176**, 181; 17

Nevada Falls 276-277, 279; 44
North America Wall 276-303, 306, 308

Onion Valley mafic complex 136; 7
Orogeny 36
Owens Valley 39, 62, 64, 67, 68, 71, 72, 100, 104, 105

Painted Lady 136, 138; 7
Palisade Creek 163-165; 14
Palisade Crest 158, 163; 13
Palisade Lake 158, 163-164; 13
Panorama Cliffs 279; 44
Paradise Granodiorite 106, 114, 116, **118**-120, 126; 5
Pegmatite 51
Pinchot Pass 94, 136, 145-146, 148-**149**, 153; 10
Pinchot pendant 149, 158; 10
Piute Creek 187-188; 22
Piute Pass 181, 188, 313, 315
Plate tectonics 34
Plucking 76, 168, 169
Pocket Meadow 198; 27
Pocket Meadow Granitoid 196; 27
Purple Lake 214, 217; 29

Quail Meadows 196; 26

Rae Lakes 133, 136; 8
Rafferty Creek 254; 39
Rainbow Falls 224-226; 32

Rainbow Falls Rhyodacite 226-228; 32
Red Cones 222-224; 31
Reds Meadow 224-225, 316
Ritter Peak 244
Ritter Range pendant 7, 58, 93, 186, 199, 216, 236-**237**, 239, 242, 246, 249, 268
Rock glaciers 86
Rosalie Lake 237; 33
Rose Lake ; 24
Rosemarie Meadow 192; 24
Round Valley Peak Granodiorite 82, 146, 197, 214, **216**-217; 28-29
Rosy-Finch shear zone 56, 193, 197, 211-214, 216-218, 311; 28-29
Royal Arches 288, 292-293, 295, 309
Ruby Lake 241-242; 34
Rush Creek 249; 36

Sally Keyes Lakes 77, 189; 23
San Joaquin River
        Middle Fork 227-228, 236; 32-34
        South Fork 183, 187, 191, 224; 21-22
Sapphire Lake 181; 18
Sawmill Pass 141, 315
Schlieren 48, 116-117, 194, 336
Selden Pass 187, 190-**191**, 199; 23
Senger Creek 188-189; 23
Shadow Creek 241; 34
Shadow Lake 238, 241-242; 34
Sheeting 44
Shepherd Pass 113, 314
Sherwin stage glaciation 83
Sierra Crest shear zone 94-95, 145, 153, 173, 237, 242, 268, 309, 310
Silver Pass 198, **199**; 27
Silver Pass Creek 198; 27
Silver Pass Granite 211; 27
Silver Pass Lake 198; 28
Squaw Lake 211, 212; 28
Subduction 34-36
Sugarloaf Granodiorite 106
Summit upland plateaus 101-104, 182
Subsummit plateaus 101-102

Sunrise Creek ; 43
Sunrise High Sierra Camp 272; 42

Taboose Pass 154, 315
Tahoe stage glaciation 84
Tawny Point 113; 4
Tenaya stage glaciation 85
Three Brothers 288, 303, 308
Tinemaha Granodiorite 141, 143
Thousand Island Lake 238, 243-246; 35
Till 33, 77
Timberline Lake 111; 1-2
Tioga stage glaciation 85
Trail Crest 110, 313; 1
Trinity Lake 236-237; 33
Tully Hole 212; 28
Tuolumne Meadows 44-45, 77, 79, 80, 236, 253, 263, 265, 270, 317; 39-40
Tuolumne Intrusive Suite 247, 248, 265, 266-269, 310
Turret Peak Quartz Monzonite 191-192; 23-24
Twin Lakes 141, 145; 10
Twin Lakes Pluton 140-141, 144-145; 10
Tyndall Creek 113-114, 118-119, 314; 4

Upper Basin 157-158; 12
Upper Bear Creek Meadows 193; 24
Upper Crater Meadow 220; 31
U-shaped valleys 127, 163, 172, 198

Vermilion Cliffs ; 27
Vernal Falls 276-277, 279; 44
Vidette Meadow 130-131; 7
Virginia Lake 214; 28
Volc. series of Purple Lake 214-217; 28-30

Wallace Creek 111; 3
Wanda Lake 176, 181; 18
Warrior Lake NA; 28
White Fork Creek 140
White Fork Pluton 137-139; 8-10

Whitney Granodiorite 109-111, 113, 114, 116, 118, 120, 127, 197; 1-5
Whitney Portal 8-9, 16, **99, 313**
Woods Creek 140, 314; 9
Wright Creek 111; 3

Made in the USA
Lexington, KY
10 September 2012